Understanding Flying Weather

Also available from A & C Black

BEGINNING GLIDING – Third Edition
Derek Piggott

GLIDING – Seventh Edition
Derek Piggott

GLIDING SAFETY
Derek Piggott

METEOROLOGY AND FLIGHT – Third Edition
A Pilot's Guide to Weather
Tom Bradbury

UNDERSTANDING GLIDING – Third Edition
Derek Piggott

Understanding Flying Weather

Second edition

Derek Piggott

A & C BLACK · LONDON

Published by Bloomsbury Publishing Plc
50 Bedford Square, London, WC1B 3DP
www.bloomsbury.com

First published by
A & C Black Publishers Ltd

First edition 1988
Reprinted 1991
Second edition 1996
Reprinted 1999, 2004, 2012

ISBN 978 0 7136 7092 9

A CIP catalogue record for this book is available from the British Library.

Printed and bound in the United Kingdom by
MPG Books, Cornwall

CONTENTS

Author's Note

My aim in writing this book has been to provide a simple introduction to Meteorology for people learning to fly gliders or other kinds of aircraft. Hopefully, the book should help the reader to understand the current weather, to interpret and make better use of forecasts, and to recognise the better days for flying and the ever present hazards associated with flying into bad weather.

Although the book is largely glider and soaring orientated, I am sure that all pilots would be safer and have more fun from their flying if they understood more of the subtleties of soaring conditions. On many days any knowledgeable pilot can use thermals and hill or wave lift to improve the climbing ability and even to increase the average cruising speed and economy of a light aircraft.

With this in mind, I have included some notes on how to make use of thermal activity when flying a light aircraft and also some advice for the less experienced glider pilot who may be having difficulties in staying up.

For a long time I have known how difficult it is for glider pilots reading for their Bronze C. It usually involves searching through a number of books, never knowing what is essential for the test. No one meteorology book seems to cover what the pilots really need to know in an easily understandable manner. I hope that this book will solve their problems and I have included questions and answers covering that syllabus. These should also be useful to the reader who is starting to study for a Private Pilot Licence.

I would like to thank John Finlater and Tom Bradbury, both very experienced forecasters, for their encouragement and for their help in checking my text and suggesting what should be included.

Derek Piggott

INTRODUCTION

Meteorology is a complex subject and therefore to simplify it means discussing only the basic factors. Each is affected and modified by the others so that all of them need to be taken into account when considering what is happening to the weather.

In trying to understand the weather from a pilot's point of view the main factors involved are:

1. where the airmass has come from and the type of terrain over which it has travelled;
2. the pressure distribution and the significance of the pressure pattern;
3. the characteristics of depressions, fronts and anticyclones;
4. the stability of the airmass;
5. the local topography;
6. the time of year and the time of day.

All of these factors and many others have to be considered by a forecaster if a reasonable prediction is to be made. However, pilots do not normally have detailed information and must rely mainly on the professional forecaster to interpret the reports and measurements available to him. Although this information is updated every few hours, many of the weather changes are rapid, making the forecasts uncertain. Even a patch of high cloud or an unexpected shower can change the soaring conditions and wreck an otherwise accurate forecast. This makes it particularly important for the soaring pilot to make a good assessment of the likely conditions by looking at the forecasts available in conjunction with his own observations.

1

THE AIRMASS

An airmass is a body of air in which horizontal changes of temperature and humidity are slight. It may extend for many hundreds of kilometres. Its characteristics are acquired by moving over a distant source region for a long period before being moved on to affect other areas. Different airmasses are separated by fronts, where horizontal changes of temperature and humidity may be sharp.

A **front** is a sloping surface separating two airmasses having a different temperature and humidity. Surprisingly, even slightly different airmasses do not easily mix together. The frontal zone where the two airmasses meet is sometimes only a few miles across, although it may on occasion stretch for 50 miles or more.

The source of the airmass determines its characteristics and in particular its temperature. However, the path over which the airmass has travelled determines how much the airmass is modified before it reaches us. Unfortunately TV and newspaper weather maps seldom give any detail on the airmasses.

Polar airmasses come from the north and start cold or cool. Cold air cannot hold much moisture and therefore polar airmasses are likely to be relatively dry.

Arctic maritime (Am). This is polar air which has only had a short sea track. The warming over the sea makes it very unstable, producing frequent showers in the north of England.

A polar airmass which has travelled over large areas of sea will be called a **Polar maritime** airmass (Pm). By originating in the north, Polar maritime air reaches us by moving southwards over a progressively warmer sea, thus warming and moistening the lower layers of air and generating instability and showery weather with good visibility.

A **returning Polar maritime** (rPm) airmass is a polar maritime airmass which has moved south over the Atlantic to a latitude south of the UK and then returned northwards round a depression

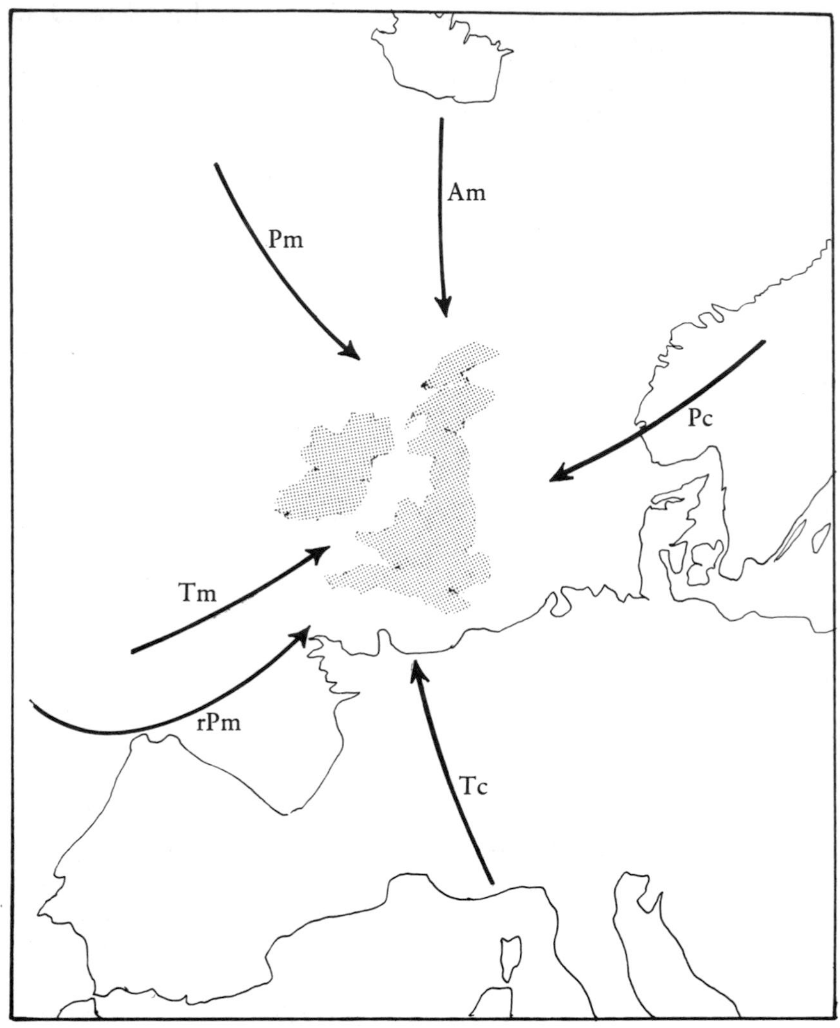

Fig 1 The origin of the airmasses. *Am* Arctic Maritime; *Pm* Polar Maritime; *Pc* Polar Continental; *Tc* Tropical Continental; *rPm* Returning Polar Maritime.

over a colder sea which stabilises the lower layers, giving stratus clouds. When this air reaches the UK it may then be heated as it moves overland. The low level stability is destroyed causing deep instability with very showery weather in the southwesterly winds.

A **Polar continental** (Pc) airmass is very cold air which comes from the east or northeast and has travelled over a large land mass. In winter the relatively warm sea may warm the lower layers resulting in cloudy, wintry conditions.

A **Tropical** airmass will be warm, but since warmer air can hold more moisture, it can be dry or have a high humidity, depending on the route it has taken to reach us.

A tropical airmass which has moved over large areas of sea will have picked up a large amount of moisture. The relatively low sea temperatures will have cooled the lower layers producing stable, cloudy conditions. Near windward coasts low cloud is liable to persist all day. This kind of airmass is then known as **Tropical maritime** (Tm) and where it travels over land in summer, the cloud will break up leaving rather stable muggy weather and poor visibility.

Where tropical air moves in from the continent in summer it will be hot, hazy and dry and known as a **Tropical continental** (Tc) airmass.

On a much smaller scale the airmass is being continuously modified during the day and night. Even a slight change in the wind direction, so that the air has travelled over more land, will often result in a dramatic change in soaring conditions. The air will be drier, giving a high cloud base and better thermals. For example, in England a small change in the wind direction of northerly winds can result in marked improvement and better soaring. Instead of blowing in almost directly from the sea, the rather moist cool air will have dried out by travelling over the length of the country before it reaches central England. This will reduce the amount of cloud cover and raise the cloud base.

With such a small island the proximity of the coastline is always a limitation for long distance glider flights. The coast line is only 20 miles or so from the Lasham Gliding Centre in Hampshire (midway between London and Southampton) and with south or southwest winds, cloud base will remain low all day and thermals will usually be weak unless good cloud streets are formed. 20 miles further inland at Booker and Dunstable the cloud base will be several thousand feet higher, giving far better conditions. Similar conditions occur near most coastlines all over the world, and have a significant effect on soaring conditions whenever the winds are coming inland off the sea, particularly in summer. In winter the sea and land temperatures are much closer to each other and sea breeze effects are insignificant.

2

THE PRESSURE PATTERN

Since the air itself has weight, the air near the surface is compressed by the weight of all the air above it. The greater the height, the less weight of air above that level and therefore the lower the pressure. Below about 5000 feet the pressure is reduced by about 1 millibar for every 30 feet of extra height. This is an average figure and is important to remember for calculating heights when using an altimeter. (*See fig.* 2.)

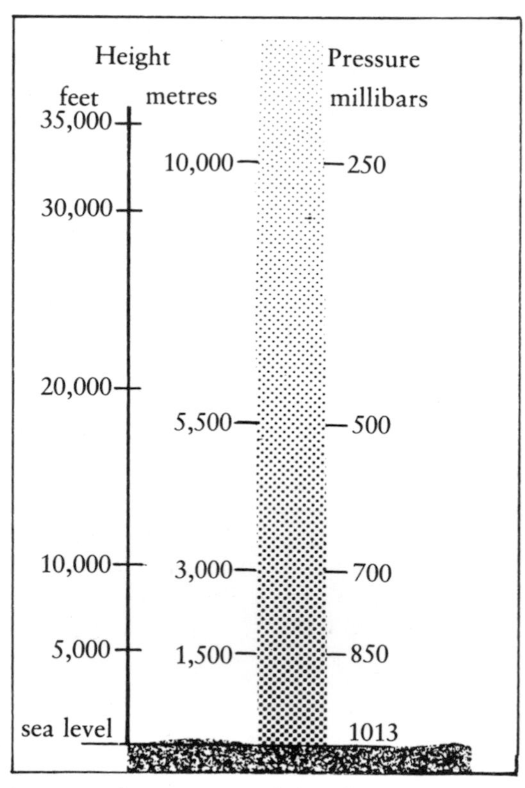

Fig 2 The reduction of pressure with height.

Lifting the air causes cooling because the atmospheric pressure decreases with height, and any reduction in pressure of a gas results in an expansion so that the same amount of heat is contained in a larger volume. This principle is demonstrated in many types of refrigerator where a gas is allowed to expand, lowering the temperature; conversely, with a bicycle pump, compressing the air into a smaller volume results in a rise in temperature.

Variations of pressure in the upper atmosphere cause a gradual ascent or subsidence in the lower levels resulting in changes in the pressure at the surface.

If air is heated, it expands when free to do so. This results in a reduction of its density and pressure. Uneven heating of the earth's surface will therefore give rise to pressure variations from place to place. Winds are the result of the movements of the air as it tries to even out these differences in pressure.

High and low pressure areas

The pressure distribution determines the direction and strength of the winds. Without the earth's rotation, air would always tend to flow directly from any higher pressure region to even out any pressure differences. (This does happen near the Equator). However, the Earth rotates and this causes air moving from high to lower pressure in the Northern hemisphere to be deflected to the right until a balance is achieved between the pressure gradient force and the deflecting force due to the rotation of the earth. This is known as the Coriolis effect, and it causes the air to flow in a clockwise direction around an anticyclone or high pressure area, and in an anticlockwise direction around a depression or low pressure area. (*See fig. 3.*) In the southern hemisphere the reverse is true and the air moves clockwise round the depressions. Since the winds are the result of these pressure patterns, to understand them and the movements of the weather systems we need to know more about pressure systems.

The readings of the local pressure are recorded by stations all over the world but it is only over the land masses that there are enough stations to form a fairly accurate picture of the ever-changing pressure patterns. In recent years the changes in pressure over the surface of the earth have been predicted largely by computer. Given the data from all the reporting stations, the computer is able to predict the most likely movements and changes in pressure patterns, enabling better forecasts to be made.

The newspaper and television charts often show the **isobars**, lines joining places of equal pressure at mean sea level (msl), and

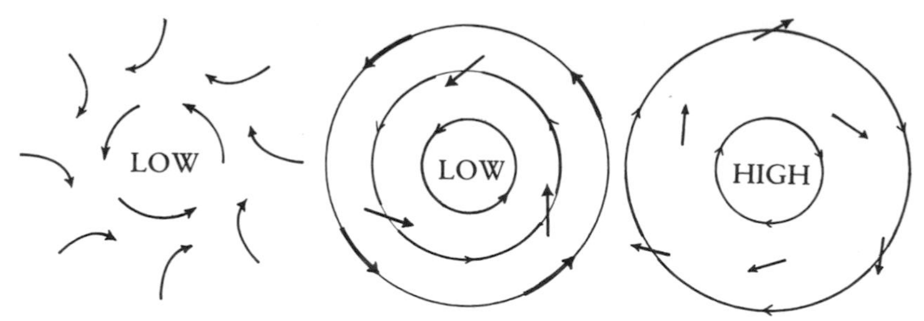

Fig 3 The earth's rotation causes the air to circulate anti-clockwise round a *low* pressure and clockwise round a *high* pressure with the surface winds at about 30 degrees to the isobars.

it is well worth while learning to interpret these charts.

It may be surprising to learn that for all practical purposes the winds at 2–3000 feet above the surface blow along these lines of equal pressure. The reason for this is not simple to explain and the average reader need only remember that this is a fact.

Near the surface, however, air is retarded by surface friction caused by trees, vegetation, structures and topography so that the wind speed and therefore the deflecting force caused by the earth's rotation are reduced. This makes the surface wind flow at an angle across the isobars towards the low pressure, often about 30 degrees overland and 10 degrees over the sea where the friction is less.

When the isobars are close together the pressure gradient and the wind are strong. When the isobars are widely spaced and the pressure gradient is weak, the winds are light.

Buys-Ballot's law

(*See fig.* 4.) This useful law states that: in the northern hemisphere, if you stand with your back to the wind, the low pressure will always be to your left and high pressure to your right. It is also useful to note that if you stand with your back to the surface wind, the upper wind at about 2000 feet will be blowing about 15-20 degrees further round in a clockwise direction.

Veering and backing

(*See fig.* 5.) The wind is said to **Veer** if it changes in a clockwise direction, and to **Back** if it changes in an anticlockwise direction. An easy way to remember these terms is that *backing* is like turning the clock *backwards*.

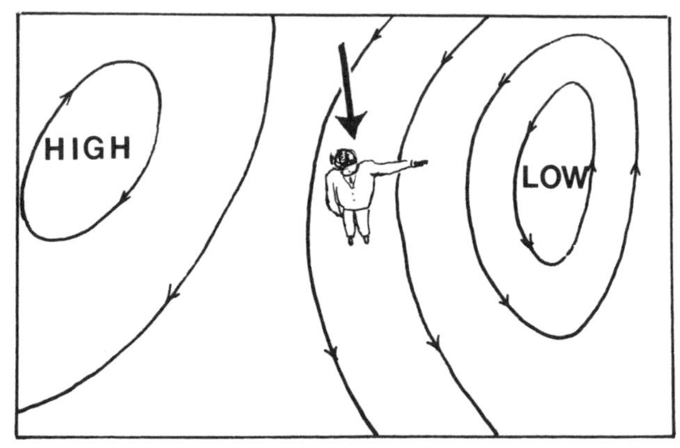

Fig 4 Buys-Ballot's Law. Standing back to the wind, low pressure is always to the left (in the northern hemisphere).

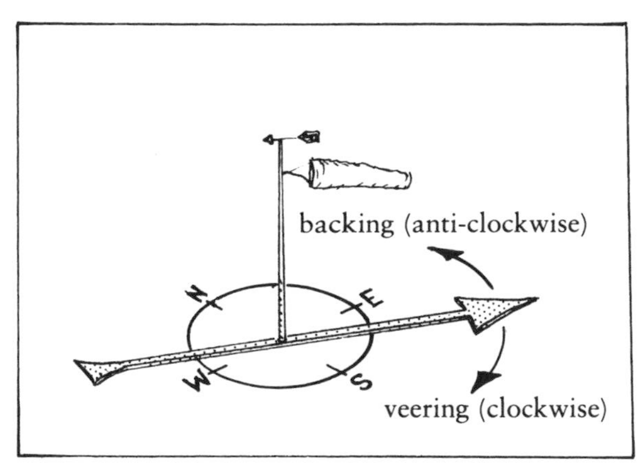

Fig 5 Veering and backing.

Interpreting the isobars

Fig. 6 (see next page) shows a weather map with only the isobars drawn in. Each line connects places with equal mean sea level pressure and in this case is marked with the pressure in millibars. The main feature is the depression centred over Northern Ireland with a weak ridge of high pressure over Europe.

Which way is the wind at Bristol? Well, the wind always blows anticlockwise round a low pressure area so we can mark in the general circulation. At Bristol, therefore, the gradient wind at 2000 feet or so would be along the isobars and in a southwest direction.

We could also predict that at the surface the wind would be

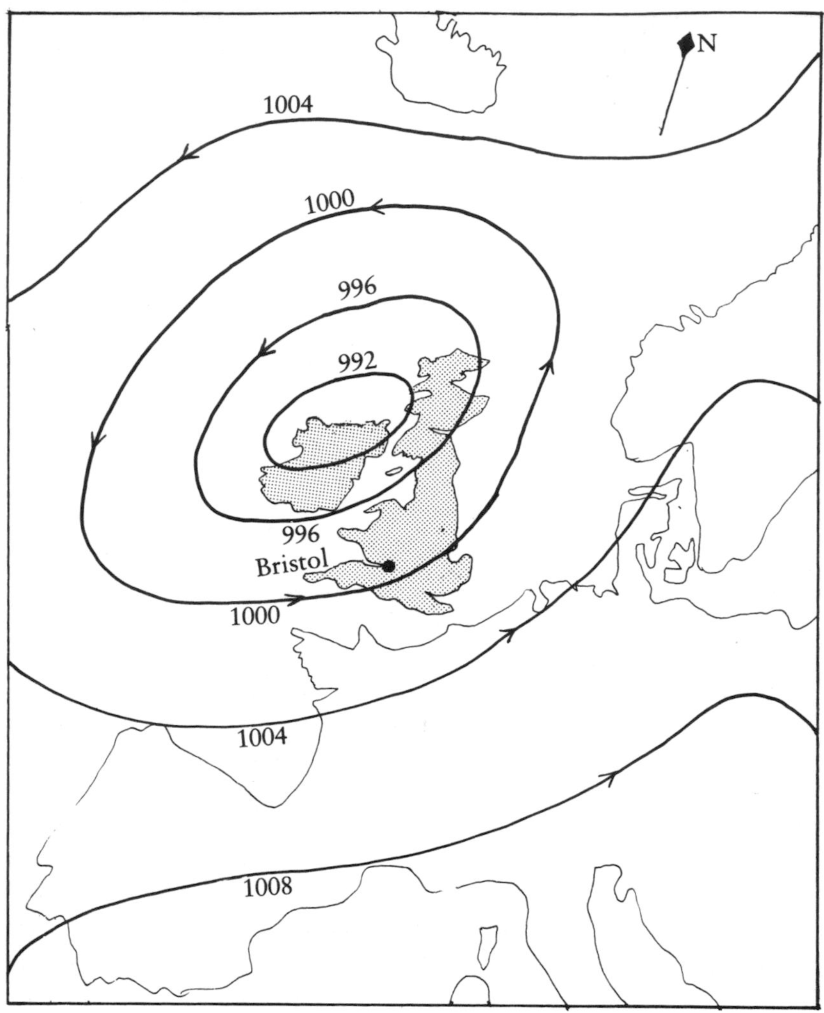

Fig 6 Interpreting the isobars. Isobars link places with equal pressure and are usually spaced at 4 millibar intervals. Which direction are the surface and upper winds at Bristol?

more southerly because it would be flowing a little more towards the centre of the low. Try the Buys-Ballot's law for Bristol. Imagine your back to the wind, Yes, the low pressure is to your left.

Wind strength

As already mentioned, where the pressure changes rapidly over a short distance, i.e. the isobars are very close together, the winds will be very strong. The Met. man can predict the wind strength

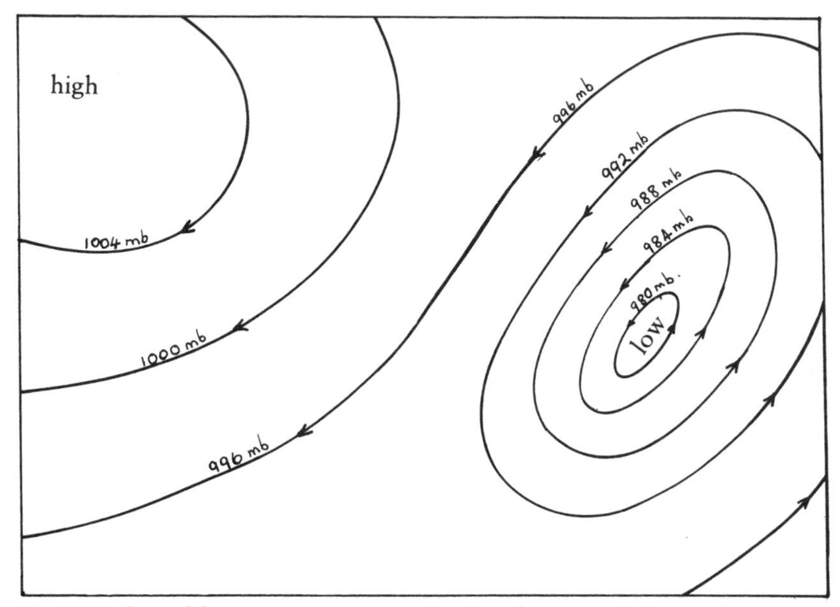

Fig 7 High and low pressure areas (anticyclones and depressions).

at a certain height by measuring this pressure gradient. A specially calibrated scale is used to measure the distance between these lines and gives him an estimate of the wind speeds at about 2000 feet. (This is known as the **geostrophic** wind.) At this height the actual wind blows along the isobars but varies from the geostrophic wind in some circumstances and this is known as the **gradient wind**. Note that this is nothing to do with vertical wind gradients which are caused by the friction of the ground on the wind and cause us so much trouble during take-off and landing in strong winds. In fact, the gradient wind is the wind *above* the influence of the surface friction. At low levels the airflow is disturbed by trees, buildings and hills, so that it is never steady. In strong winds, gusts of at least the strength of the upper wind should be expected because some of the momentum of that upper wind is brought down by the effects of convection and turbulence.

Fig. 7 shows a low and high pressure region. In the high pressure area (an **anticyclone**) the isobars are further apart indicating much lighter winds. Winds always blow clockwise round a high.

It is usual to have a much lighter surface wind in the early morning and late evening than during the middle of the day because the cold dense surface air increases the friction effect. When the surface air warms and thermal activity begins it mixes the surface wind with the upper winds. Some air is taken up and some brought down, with the result that the lower winds increase and

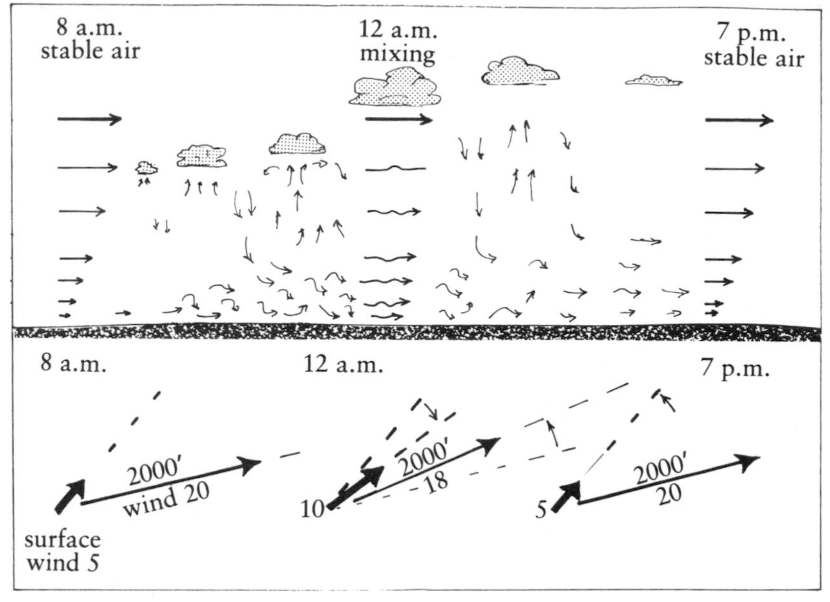

Fig 8 Changes in the surface and upper winds during the day. The surface winds increase and veer as the thermal activity causes mixing, and decrease and back in the evenings as the air becomes stable.

swing a little closer in direction to the wind at 2000 feet or so. (*See fig. 8.*) A knowledgeable Duty Instructor at a gliding site will allow for this inevitable veering and strengthening of the surface wind when he sites the position of the winches in the early morning.

For example, if the surface wind is a light westerly at 8a.m., it will most likely increase by 5 knots or so and swing to a more northwesterly direction later in the morning. However, this effect can be completely swamped by any rapid change in the weather, such as the approach of a depression.

The surface wind may also increase and change in direction if it is reinforced by the sea breeze. The timing of this increase will depend on the distance the air has to come from the coast. For example, about 25 miles from the coastline an onshore wind can be expected to increase sometime after mid-day, bringing in cooler, moister air and probably spoiling any thermal activity. Sea breezes may also advance inland against a generally light but opposing wind, resulting in changes in the wind direction as well as in strength.

During the daytime the high ground heats up more rapidly than the valleys and this sets up winds, known as **anabatic** winds, blowing up the slopes. (*See fig. 9.*) This flow up the mountainsides is confined to quite a shallow layer so that using it involves flying

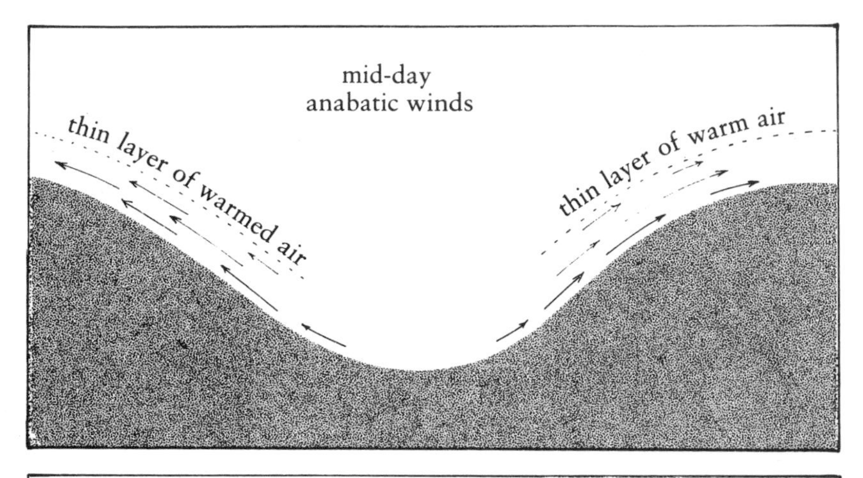

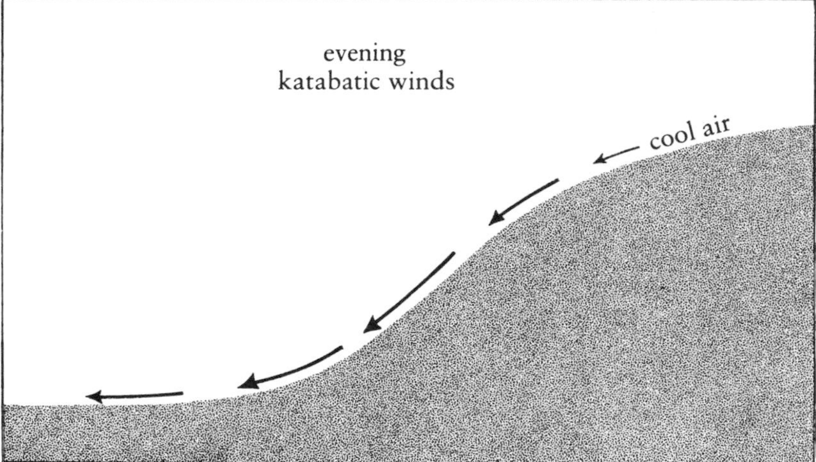

Fig 9 Anabatic and katabatic winds.

very close to the hill slope. Anabatic winds are most evident on sun-facing and lee slopes where the surface gets the greatest heating.

Conversely, in the evening and particularly on a clear night, the earth radiates heat out into space, cooling the ground and the air close to it. In hilly country this cooler air flows down the sides of any hills or mountains into the valleys. This is known as a **katabatic** wind.

Low-level winds will always tend to blow round obstructions such as hills, causing quite large variations in both the direction and strength near the surface.

In the lee of hills and mountains the surface wind strength will vary considerably if lee waves are formed. Below the crests of the waves the winds may be very light or even reversed, whereas only a short distance up or downwind of these areas the wind can be very strong indeed. Even if it is a light wind on the ground precautions always need to be taken to tie down gliders properly, as the position of the wave may change without warning, resulting in a sudden unexpectedly strong wind. (Lee waves are explained in a later chapter.)

Wind gradient

(*See fig. 10.*) The layers of air close to the ground are slowed down by the friction between the air and the ground. Obviously, the stronger the wind, the greater the change in the wind speed which is possible due to this effect. The friction will also be greater if the ground is rough or covered by obstructions. If the air is unstable and the area is in the lee of obstructions such as trees or buildings, some of the faster moving air from above may be brought down to ground level, causing momentary stronger winds or gusts. These rapid changes in wind speed combined with the wind gradient often make take-offs and landings critical in windy weather. Flying down into a rapidly decreasing wind speed during the final stages of an approach causes a sudden loss of airspeed and a rapid loss of height which can be dangerous in a glider or light aircraft. This effect is caused by the **wind gradient**. Turbulence near the ground always indicates rapid changes in wind speed and direction and extra flying speed is essential to ensure that there is adequate speed and control at all times.

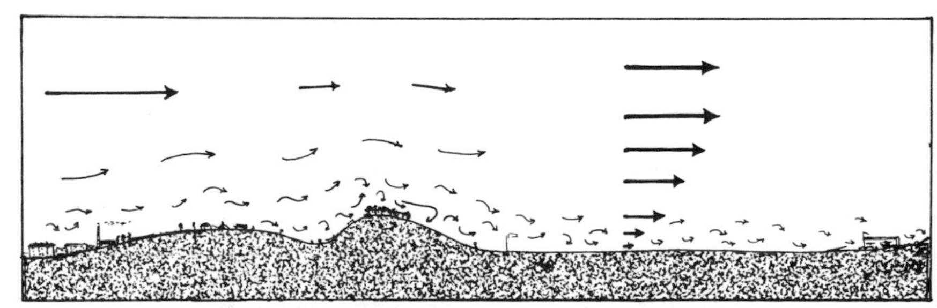

Fig 10 The wind gradient. Surface friction reduces the wind speed near the ground.

3

DEPRESSIONS AND ANTICYCLONES

In the past few years, with the developments in radar and the use of powerful computers and weather satellites, the mechanisms by which the depressions develop have been studied in much more detail than was ever possible in the past.

It is now known that movements occurring in the upper atmosphere are the cause of the high and low pressure systems forming near the surface which dominate our weather.

With the air circulating round the low pressure area, air must be rising and being removed at high level or the **low** would fill in very quickly. Similarly, for a **high** pressure system to persist there has to be a descent of the air to maintain the higher pressure. It is these general movements up and down, started at high level, which dictate where systems will form and which cause the characteristic weather associated with them.

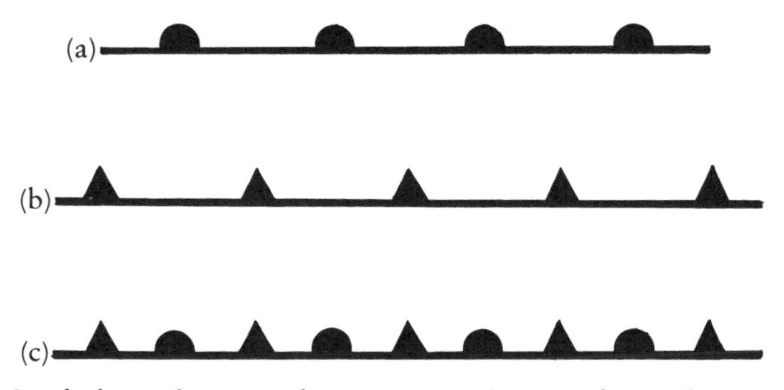

Fig 11 Symbols used on weather maps. (a) A warm front. (b) A cold front. (c) An occlusion.

Depressions

The weather systems moving into Europe originate in the Atlantic and generally move from west to east. The cold airmass from the polar regions meets the warmer tropical air along a **front** and depressions form along this boundary.

It is known as a **warm front** if the warmer air mass is moving forward over the surface, and a **cold front** if it is the cooler air mass moving forward over the surface. In some circumstances the movement of a front may be reversed so that a warm front becomes a cold one.

The majority of our weather systems form out in the Atlantic along the frontal zone between the Arctic air mass and the warmer Tropical maritime air to the south.

When two air masses of different density lie side by side they induce a strong current of air to flow parallel and on the cold air side of the front at very high altitudes. This is known as a **jet stream** and is several miles deep and travels at speeds of 100 to 200 miles an hour. (*See fig. 12(a).*)

Disturbances in this upper flow cause the jet stream to change direction and swing from side to side and this results in areas where the flow is speeded up or slowed down. (*See figs 12(b) and (c).*) Where the flow speeds up because of divergence, the pressure drops slightly inducing an upward flow of air.

This continuing upward movement results in a change of pressure at the surface which forms the depression. During the development of a depression air is flowing in near the surface, though not fast enough to stop the depression deepening. (*See fig. 13(b).*)

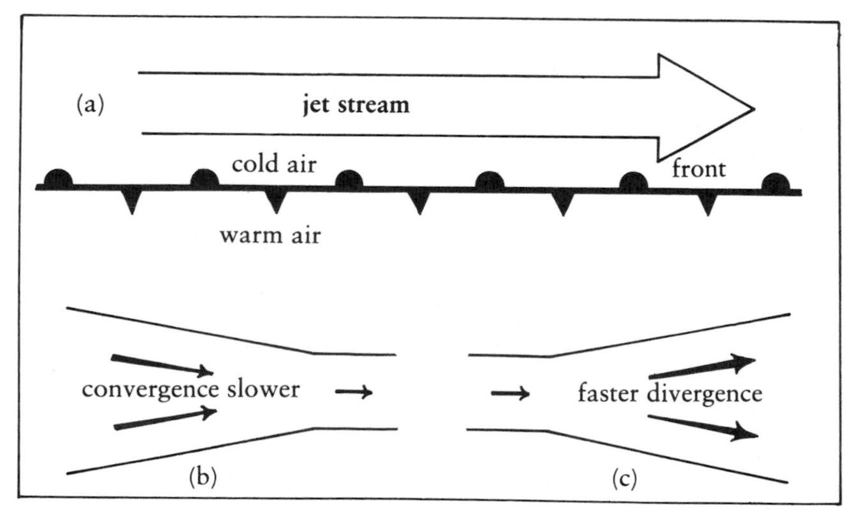

Fig 12 (a) Upper airflow along a front. (b) Convergence. As the speed of flow decreases, the pressure rises. (c) Divergence. As the speed of flow increases, the pressure decreases.

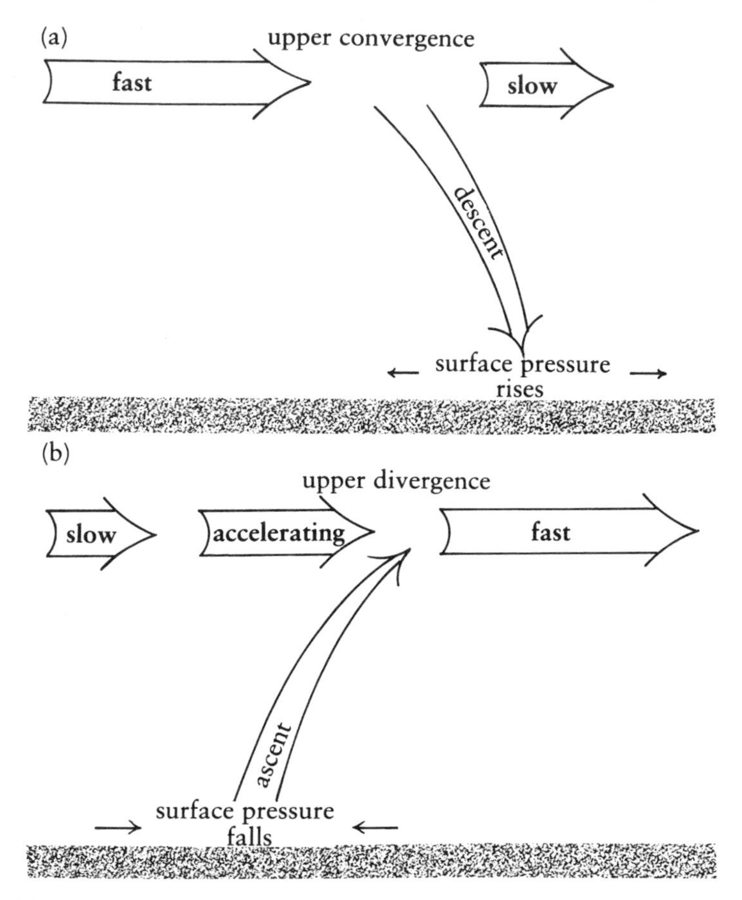

Fig 13 (a) Upper convergence causes an increase in pressure and a descending flow forming an area of high pressure (anticyclone). (b) Upper divergence causes a drop in pressure and an upward flow, reducing the surface pressure and forming an area of low pressure (depression).

Similarly, anticyclones start with convergent flow slowing down the airflow at height and causing an increase in pressure and a downward movement of air. This results in an area of higher pressure and an outward flow at low levels. (*See fig. 13(a).*)

The life of a depression

In figs 14(a) and (b) the increase in the speed of the upper flow caused by a divergence of the air at height has reduced the pressure. This starts a flow upwards, reducing the pressure at the surface, and induces an inflow of air trying to fill in the area of lower pressure.

(*See figs 14(c) and (d).*) Because of the effects of the rotation of the earth, this inflow is diverted into a circulation in an anti-clockwise direction.

Whereas at first the frontal surface was more or less stationary, the circulation starts a wave in the front. The cold dense air moves and starts to undercut the warmer tropical air forming a **cold front.** At the same time the warmer air is moved up over the colder air mass forming a **warm front.** (d)

(*See figs 14(e) and (f).*) As the air is continuously extracted at the top of the system, so the pressure at the surface continues to drop, increasing the circulation and the wind speeds. The anticlockwise circulation is gradually spread up to higher levels and the upward movement of the warmer moist air causes adiabatic cooling and the formation of clouds, releasing more heat energy as the moisture condenses into water droplets.

The cold front moves more quickly than the warm one and catches it up forming an **occlusion,** with the **warm sector** at the surface being reduced in size.

See figs 14(g) and (h). The warm air continues to be carried round in a spiral by the winds but the upper extractor effect has ceased, so that the circulation runs down with the surface low pressure area gradually filling. However, the changes in the upper flow caused by the last stages of the life of this depression often create a further acceleration of the flow in the jet stream. This sets off the process again, forming a new depression.

Sometimes a whole family of depressions forms, giving a period of almost continuous unsettled or bad weather. On other occasions the upper flow will be slowed down, causing an increase in pressure and a gradual descending flow which starts the formation of an anticyclone and more settled weather.

The meandering of the jet stream to the north and south as it flows across the Atlantic largely determines the pattern of our weather and the position and intensity of the areas of high and low pressure.

Warm and cold fronts

Both warm and cold fronts may vary in the weather conditions they produce, and may differ considerably from the idealised examples shown in the illustrations. The slope of the front will vary, causing changes in the cloud and the amounts and intensity of the rainfall along the front. Some warm and cold fronts may even have gradually descending air and this will reduce the amount of activity on them. Some warm fronts will have active

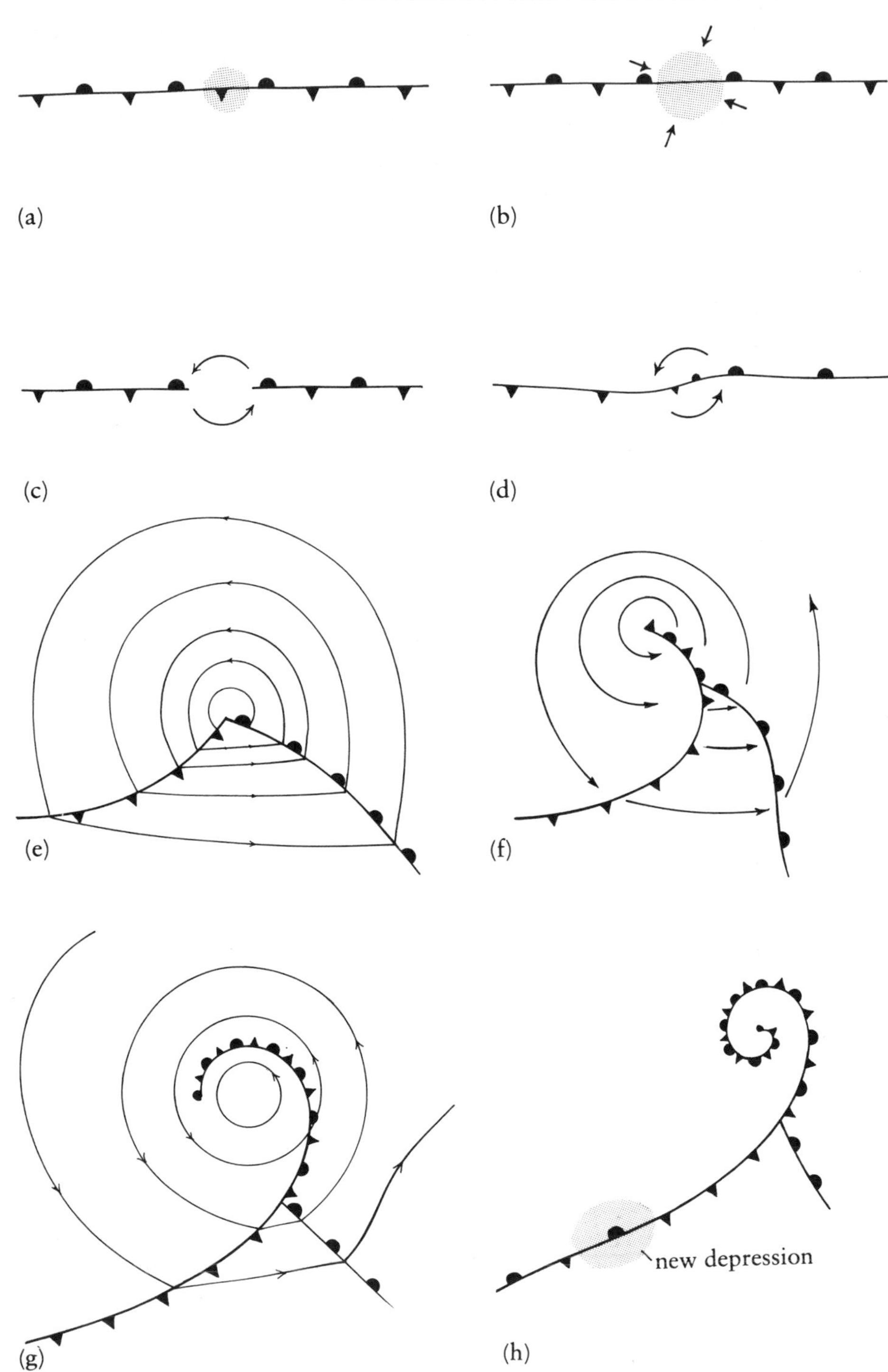

Fig 14 The life of a depression.

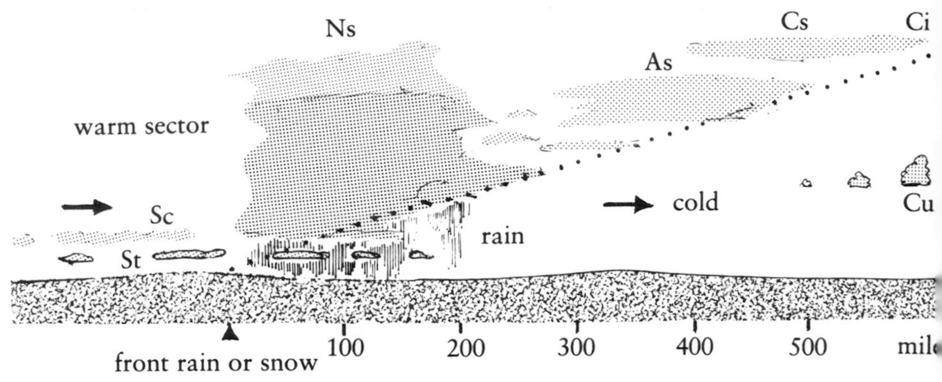

Fig 15 A section through one type of warm front.

thunderstorms and many of the characteristics of a cold front. Because of the many variations in fronts and depressions, I have not attempted to explain all the types of front.

It is important to realise that it is very rare indeed to have both an active warm front and an active cold front on the same depression. This means that if there is a large amount of rainfall on the warm front there will be little activity on the cold one.

The normal warm front has a shallow slope with the air rising very gradually over a distance of many hundreds of miles. This lifting produces the gradually thickening layer of cloud which results in steady rain near the front. (*See fig. 15.*)

Usually the high cloud cuts off the sun's heating, but if the high cloud arrives after the thermal activity has become well established, it is not uncommom for soaring conditions to persist for several hours after the sun has become partially obscured.

The warm front drizzle and rain sometimes start sporadically, but once the cloud has thickened and continuous rain begins, it is usually the end of flying for many hours, if not for the whole day.

Fig. 16 shows a typical depression with warm and cold fronts and their probable rain areas. In reality only one of the fronts is likely to be active. The first indication of the approaching bad weather will be the high cirrus cloud, possibly creating a halo around the sun or moon. The cloud will gradually thicken to altostratus and nimbostratus, with rain beginning 5–10 hours before the passage of the warm front. At this stage the winds will usually strengthen and back to become more southerly as the pressure falls. At the front the rain will ease off, the wind will veer 50 degrees or so, and the temperature and humidity will rise noticeably. The warmer, muggy weather is usually quite obvious after the front has passed. On the weather map the distinct kink

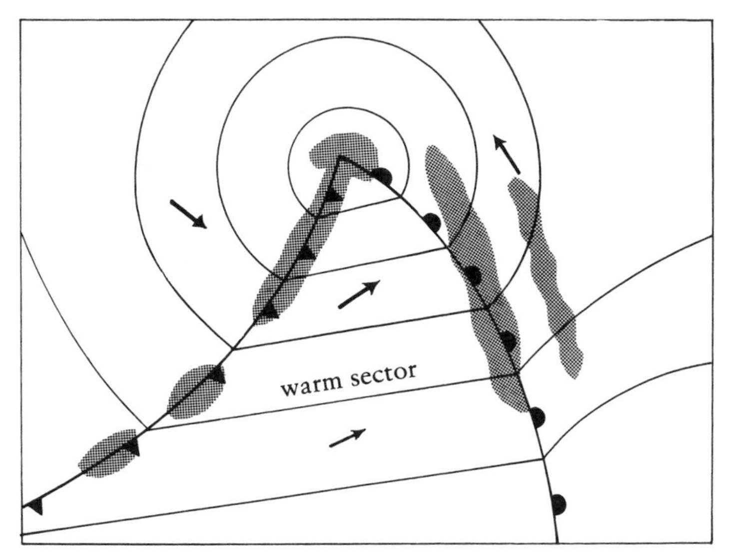

Fig 16 Warm and cold fronts on a depression, showing areas of probable rain and the surface winds.

in the isobars at the front indicates the change in wind direction.

The **warm sector** between the warm and cold fronts is typified by relatively warm humid conditions and low cloud. It may produce drizzle, especially in winter, but there are often good breaks in the cloud allowing training flights to continue although there will be few if any thermals. The warm sector may also have suitable conditions for usable waves to form in the lee of hills and mountains. (Lee waves are explained later.)

Notice that the cold front is a relatively steep wedge of air shovelling up the warmer air rapidly. (*See fig. 17.*) The passage of

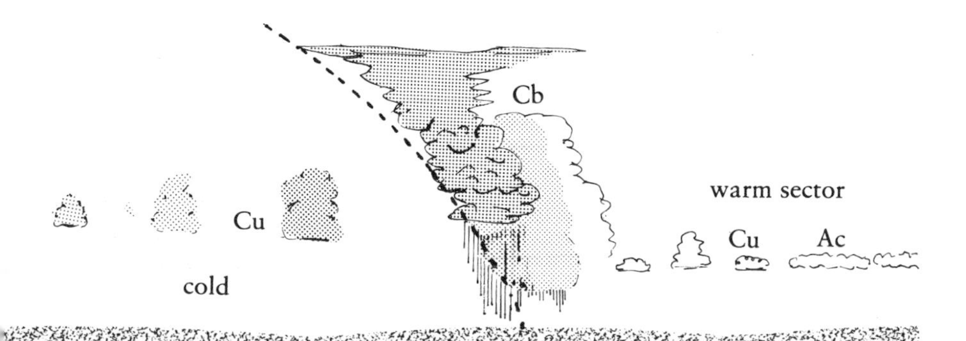

Fig 17 A section through one type of cold front.

Above A squall line below an active cumulonimbus. (Photograph courtesy of John Simpson)

Below An active sea breeze front, showing the double cloud base and wispy 'curtain' clouds. (Photograph courtesy of Lorna Minton)

an active cold front is often quite dramatic with heavy showers, thunderstorms and squally winds. At the front the temperature drops, the air is drier and again the wind veers a further 60 degrees or so, usually to a northwesterly direction. Behind the cold front there is often a sudden complete clearance and a period with little or no cloud. Alternatively the cold unstable air behind the cold front may encourage the development of further showers.

In summer, a line of thunderstorms may develop far ahead of a surface cold front during the afternoon. These may produce a fierce and unexpected **line squall**. In this case the actual cold front will only produce light rain and low cloud. So, in summer it is important to watch for signs of this happening long before the cold front is due as it can produce very severe conditions.

Particularly with the passage of a cold front, the weather can change completely in a few minutes. It has been known for the morning briefing for the National Championships to be held to the accompaniment of heavy rain and very low cloud, with the competitors unable to believe the forecast for a clearance at midday. Then, as predicted, the skies have cleared, the sun has come out and within an hour or so all the gliders have been launched for a successful cross-country task. Perhaps surprisingly, if the airmass is very dry and cold, good soaring conditions can occur even after the heavy rain on the front.

After the passage of a cold front, the approaching high pressure gives us the northerly winds bringing air down the whole length of England and producing the very best days for soaring. In most European countries similar weather patterns also produce ideal conditions for long flights. This is usually on the day after the passage of a cold front. The ground will have had time to dry out, and with the ridge of higher pressure moving in, the air will usually become more stable, preventing showers and leaving a limited depth of thermal activity. With little or no high cloud and only small cumulus clouds forming, these are the days for long distance cross-country flights in gliders. When the pressure begins to fall again, as the high pressure moves away to the East, the conditions deteriorate and any cloud tends to persist and cut off the sun's heating.

Occlusions

As the depression deepens, the cold front gradually overtakes the warm one so that the weather near the fronts becomes confused. The heavy showers and thunderstorms may then be concealed amongst the deep layer clouds of the warm front. This is known

An ideal soaring day. Cumulus cloud forming over high ground first.

By afternoon the cloud base is much higher than in the morning and is regularly spaced as far as the eye can see.

as an **occlusion** or an **occluded front** and is indicated on weather maps by a line with alternate warm and cold front symbols.

Sometimes the cold front will undercut both the warm sector and the cooler air beyond. If the cold front air happens to be slightly warmer than the cold air below the warm front it will ride up over it. (*See fig. 18.*) Both these types of occluded fronts have a mix-up of weather and slightly different characteristics. Both, if still active, involve bad, non-gliding weather.

Eventually all the air in the warm sector has been lifted up by

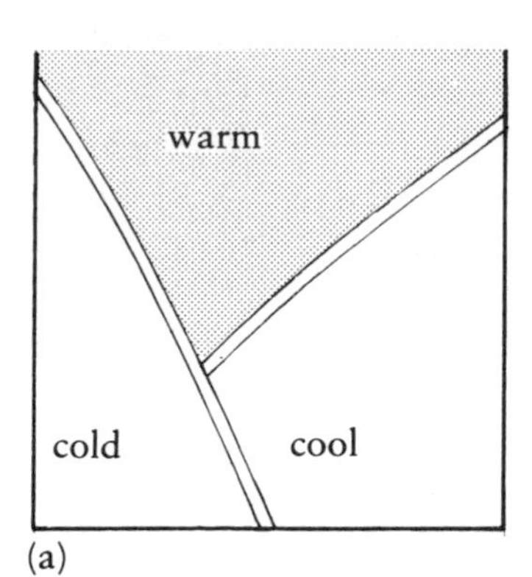

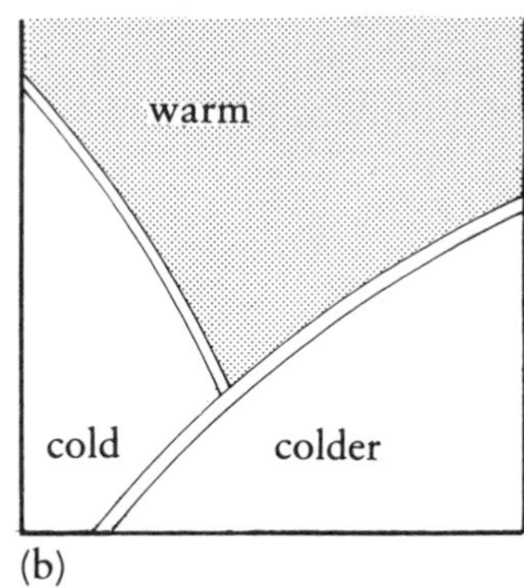

Fig 18 Types of occlusion. (a) Cold air undercutting the warm front. (b) Cold air riding up over the original warm front.

the warm and cold fronts, leaving only a pool of slightly warm air aloft with some high level cloud. Slow-moving occlusions are a common cause of persistent periods of continuous rain and low cloud.

Not every depression brings very bad weather. A depression may arrive at our shores at any stage of development; it may have an active or weak warm and cold front, or it may have no fronts or a decaying occlusion. There may be very little to indicate the passage of the fronts except for the change in wind direction and humidity. The weather experienced during a depression may vary from some cloud with little or no rain, to some hours of continuous rain or a vicious line squall and severe thunderstorms. It is comforting to note that if there is heavy rain on the warm front the following cold front is unlikely to be very active.

An active cold front must always be taken seriously as it may produce violently squally winds, torrential rain, hail and lightning hazards. Gliders must *never* be left out unattended if an active cold front is forecast and even powered aircraft should be securely tied down or hangared.

Unexpectedly strong winds and severe weather will develop with very little warning if a small secondary depression forms on the cold front. This starts as a 'wave' on the cold front and is particularly bad news. Instead of the normal rise in pressure as the depression moves to the east, the secondary low forms very rapidly, sweeping round on the end of the old cold front and producing severe gales and storms. Sometimes a whole series of little wave depressions may form along the trailing cold front, producing several days of bad weather until another high pressure system moves in.

Another source of an unexpected change in the weather is a thundery depression. This is a small but violent disturbance forming during a heat-wave and resulting in thunderstorms.

Anticyclones

As already mentioned, the lowering pressure in the depression occurs with the general lifting of the air, creating cloud and bad weather. Conversely, in an anticyclone, the air is gradually subsiding and being warmed as it is compressed. Since the warmer air can hold much more moisture any cloud tends to disperse. This very gradual subsidence results in clear skies at night and little tendency for cloud to over-develop and cut off the heating from the sun during the daytime. In a developing anticyclone this subsidence is about 20–40 metres per hour (about a quarter of an inch per second) but it is much less in a fully developed system. This is too small to be detectable by instruments when flying, though it would result in a significant error when testing the performance of

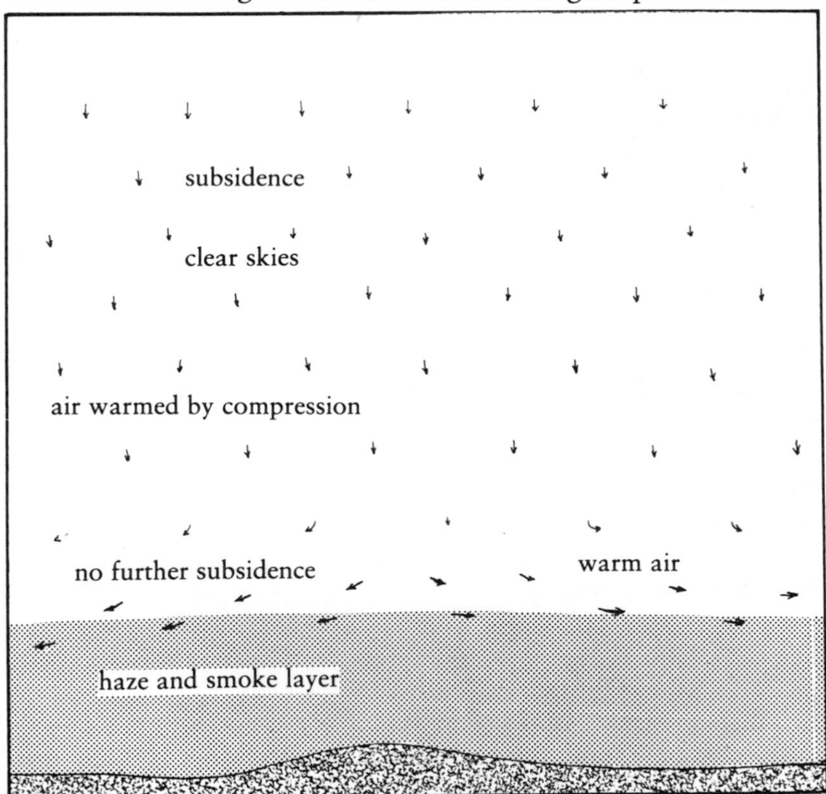

Fig 19 The gradual subsidence in an anticyclone results in an inversion trapping any smoke and haze.

a modern glider. Here an extra half an inch per second or so would reduce the gliding angle by several points.

In the summer the approach of higher pressure always means an improvement in the weather with lighter winds and less cloud. However, in winter a high pressure system can mean persistent fog and low cloud or it may lead to clear skies, depending on the source and track of the air at low level.

High pressure systems move more slowly and are more persistent than depressions. Near the surface the air is no longer descending and being heated by compression. Instead there is an outward flow so that this air remains relatively cool and moist. With cool air at low levels and warmer air above there is an increase in temperature with height which is known as an **inversion**. The atmosphere becomes stable, restricting any thermal activity. In addition, the clear skies at night allow the earth to radiate its heat into space, cooling the lower layers of the air even more. (*See fig. 19.*) If the anticyclone persists for three or four days, the weather may be good for sunbathing, but in spite of the sunshine and heat will be useless for soaring because of the intense, rather low level inversion. The weather will become more and more hazy because any dust or smoke will be trapped below the inversion.

A stationary high pressure system over the Continent will often block the progress of a depression moving eastwards and make it move northwards over Scotland. Sometimes it may even force it back so that the bad weather persists over a period of days instead of moving off into Europe.

There are various other types of pressure systems mentioned in weather forecasts. These are illustrated in Fig. 20.

A **ridge** of high pressure is a wedge-shaped area of high pressure between two areas of low pressure.

A **trough** of low pressure is an elongated area of low pressure without any distinctive frontal systems but usually having rain and cloudy weather.

A **col** is an area of slack pressure between two lows and two areas of higher pressure.

Timing

Although the basic forecast of the approach of a depression is unlikely to be wrong, the timing of the bad weather is difficult to predict and may be many hours out. This is where an observer such as the individual pilot can often recognise whether the forecast is likely to be correct.

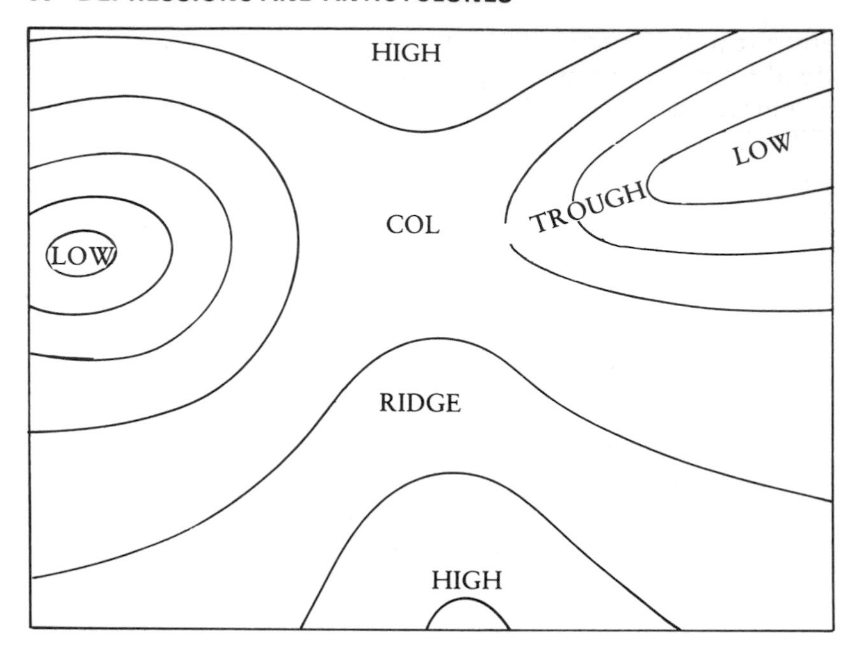

Fig 20 Common pressure systems: A *ridge* of high pressure; a *trough* of low pressure; a *col*.

Normally a depression will move at about the speed and direction of the gradient wind in the warm sector (if there is one), perhaps 15 to 20 knots. If there are no fronts and no warm sector the low will move in the direction of the strongest winds indicated by the closer isobars. However, an approaching low may be held off for many hours by an unexpected rise in the pressure, turning the forecast of deteriorating weather into a good soaring day. Even quite a small rise in pressure can mean a significant improvement in the weather.

The rate at which the pressure falls is a good guide to the speed at which the weather will deteriorate. As a depression approaches, the cloud base will lower and the wind can be expected to back and strengthen just before the rain starts.

An error in the timing of the arrival of the bad weather often occurred in the past but is now less likely with computer forecasting. Its arrival can frequently be anticipated by checking the shipping reports and the **volmet** broadcasts which are actual observations updated every hour. If the rain is already falling on the west coast and is moving at 15–20 knots it is easy to estimate when it is likely to arrive in your location. On other occasions the pressure may be reported to be rising slightly instead of showing the expected fall, and this means that any bad weather is likely to be delayed. Small rises in pressure to the west, together with no sign of the edge of the high cloud from the approaching depression, would encourage most pundits to rush to their gliding site and get rigged ready to fly.

The TV forecasts are particularly useful, as the satellite pictures give a good indication of the positions of any bad weather systems and of what changes are taking place.

The VOLMET is continously broadcast on the VHF air band for the airlines on: London (North) 126.60 mHz, London (South) 128.60 mHz, and London (Main) on 135.375 mHz. VOLMET includes the surface wind, temperature and dew point, cloud base and the amount of cloud, for the principal airports in the UK.

The RAF have their own VOLMET giving actuals for Service stations in other locations and this information can be very useful in determining the flying prospects for the day. This is on SSB (single side band) on 4722 and 11200 kHz and requires a special radio to receive it.

Shipping Forecasts

In order to make the best use of these weather bulletins you need an enlarged copy of Figs. 21 and 22 covered with perspex or a similar plastic so that you can note the detail against the appropriate area or place. This should give you a more complete picture of the situation.

Shipping forecasts are on BBC Radio 4 200 kHz (1500 m) at the following times GMT:

00.33 to 00.38
05.55 to 06.00
13.55 to 14.00
17.50 to 17.55

They consist of the following details:

1 gale warnings;
2 general synopsis for the next 24 hours giving the positions and movements of areas of high and low pressure;
3 the forecast for each sea area giving the wind velocity, weather and visibility;
4 the latest reports from some of the coastal stations.

The forecasts for the coastal sea areas are broadcast in the following order. Of course on a larger map you will be able to print in the names.

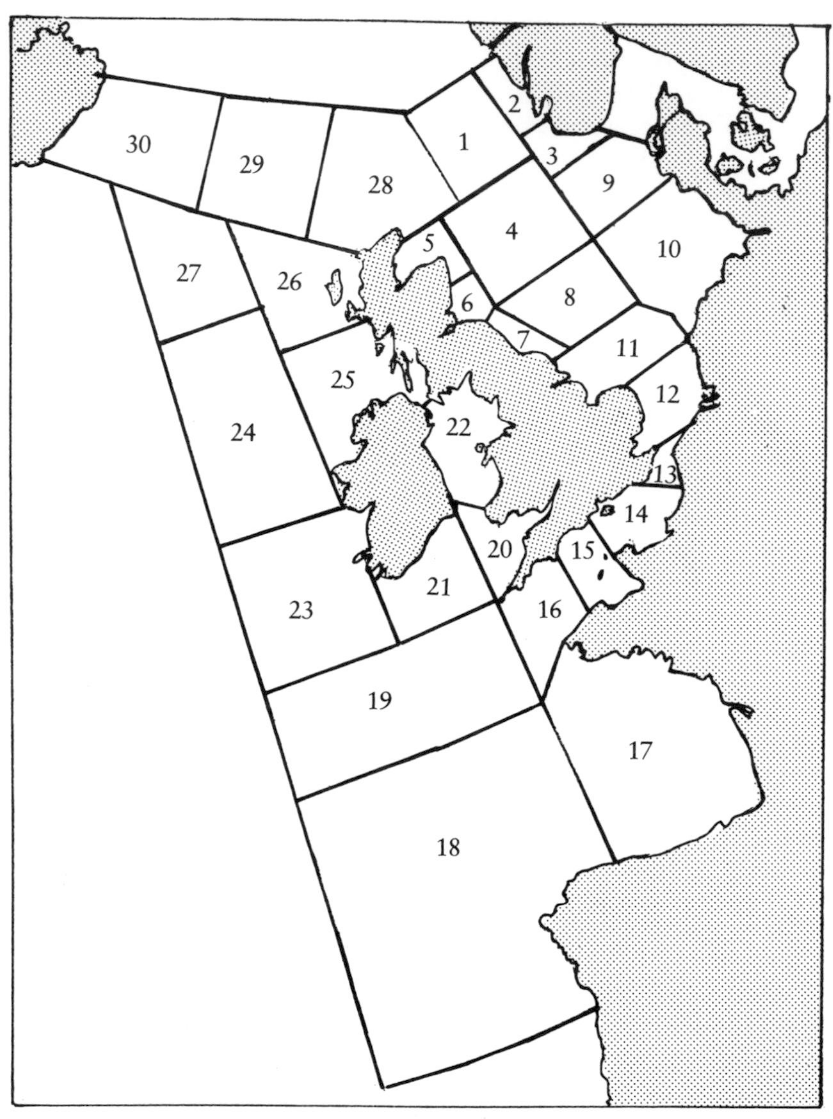

Fig 21 Sea areas used for BBC Shipping Forecasts.

1 Viking, *2* North Utsire, *3* South Utsire, *4* Forties, *5* Cromarty, *6* Forth, *7* Tyne, *8* Dogger, *9* Fisher, *10* German Bight, *11* Humber, *12* Thames, *13* Dover, *14* Wight, *15* Portland, *16* Plymouth, *17* Biscay, *18* Finisterre, *19* Sole, *20* Lundy, *21* Fastnet, *22* Irish Sea, *23* Shannon, *24* Rockall, *25* Malin, *26* Hebrides, *27* Bailey, *28* Fair Isle, *29* Faeroes, *30* South East Iceland.

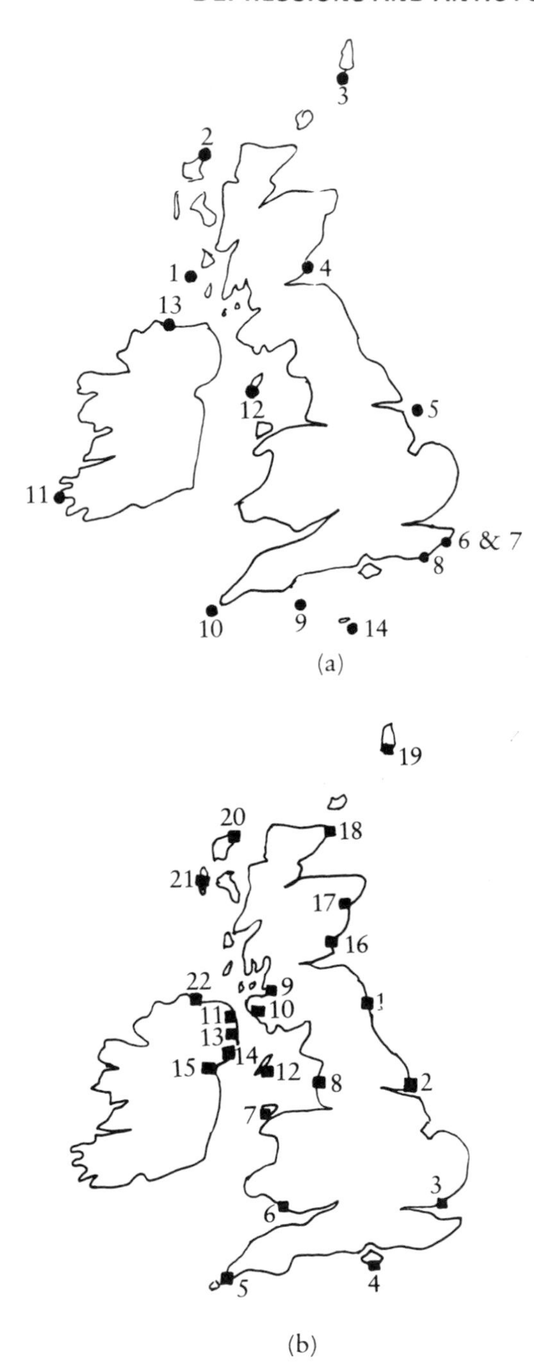

Fig 22 (a) BBC coastal station reports. (b) BBC forecasts for inshore waters. (See next page for key.)

The latest reports from coastal stations

These report the wind velocity, present weather (including that for the past hour), visibility, pressure and pressure tendencies. They are given after the sea areas forecast, in the following order:

1 Tiree, *2* Butt of Lewis, *3* Sumburgh, *4* Bell Rock, *5* Dowsing, *6* Dover, *7* Varne, *8* Royal Sovereign, *9* Channel Lightvessel, *10* Scilly, *11* Valentia, *12* Ronaldsway, *13* Malin Head, Jersey. (*See fig. 22(a).*)

Forecasts for inshore waters

These are broadcast on BBC Radio 4 at the end of the evening programmes at midnight and are valid until 18.00 hours the next day. The forecasts of wind, weather and visibility are followed by reports giving the wind velocity, present weather, visibility, sea level pressure and tendency from the following stations:

1 Boulmer, *2* Spurn Point, *3* Walton-on-the-Naze, *4* St. Catherine's Point, *5* Land's End, *6* Mumbles, *7* Valley, *8* Blackpool, *9* Prestwick, *10* Corsewall Point, *11* Larne, *12* Ronaldsway, *13* Orlock Head, *14* Killough, *15* Kilkeel, *16* Leuchars, *17* Aberdeen (Dyce), *18* Wick, *19* Lerwick, *20* Stornoway, *21* Benbecula, *22* Malin Head.

Perhaps not many pilots want to stay up to hear the midnight shipping reports, but when the previous day's forecast looks promising, the early morning one can often give a useful clue to whether the earlier forecast is going to come up to expectations or whether the next depression has moved in more quickly than was expected.

The 'Breakfast Time' TV weather forecast is the easiest to understand but does not necessarily tell the same story as the BBC Radio broadcasts. The forecaster has the same basic information but makes the interpretation himself.

4

THE ATMOSPHERE

Water vapour

Water vapour is an invisible gas which is actually a little lighter than air. When we look at a boiling kettle, at first invisible gas comes out of the spout. After a short distance the rapid cooling causes the water vapour to condense into the visible water droplets we know as steam. However, because the air around is relatively dry the steam evaporates a few inches further on. This is really cloud making on a small scale. If the air is very moist, as in the bathroom, the water droplets take longer to evaporate. This makes the bathroom 'misty' and the water vapour will condense out as liquid on any cold surface such as a mirror or window.

Whereas water vapour is an invisible gas which is lighter than air, clouds formed of water droplets and ice particles are visible and much heavier. With cumulus clouds the water droplets start to evaporate in the drier air surrounding the cloud, causing extra cooling. This causes the air to sink, and since sinking air is compressed and warmed by the increase in atmospheric pressure, the evaporation of the water droplets accelerates. This process makes the life of any individual cumulus cloud very short unless it is being continuously fed by new thermals.

If the cloud forms at a greater height, or the water droplets are carried up above the freezing level, they become **supercooled**. Supercooled water droplets can remain unfrozen to very low temperatures (–30°C in some circumstances) provided that there is no disturbance in the air. To form ice crystals the supercooled droplets need to freeze onto some kind of nuclei such as dust, clay or soil particles. High-level cirrus clouds are composed of very small ice crystals and these fall more slowly and take much longer to evaporate than water droplets. The 'anvil' tops of cumulonimbus storm clouds which are composed of ice crystals take a long time to disperse. To the dismay of the soaring pilot, the sun's heating

Large cumulus cloud developing.

The top has become ice crystals spreading out into the anvil, an active cumulonimbus cloud. (Photographs courtesy of Chris Bryant)

can be cut off for many hours by even a thin layer of cirrus cloud after a heavy shower or thunderstorm.

Showers will occur when the cloud beings to **glaciate** or change into ice crystals. Once the supercooled moisture starts to freeze, the water vapour quickly condenses out to enlarge the crystals which melt and turn to rain. Heavy showers of rain involve the water droplets being carried up in the cloud to great heights to allow them to grow larger. For showers to occur in this country it is usual for the cloud to have to extend to above the freezing level. Thus the probability of showers developing can be judged by checking the forecast height of the cumulus tops in relation to the freezing level. (If the cloud tops are high and the freezing level is low, it will probably be a very wet day.)

Stability and instability of the atmosphere

In an ideal situation, with no heating from the sun and no cooling at night, the temperature of the air would decrease regularly with an increase in height. This reduction in temperature is called a **lapse rate** and where the air is not saturated and forming cloud, the lapse rate is a regular 1°C (or, to be more precise, 0.98°C) per 100 metres gain of height. This is known as the **Dry Adiabatic Lapse Rate** or DALR.

An adiabatic process is one where heat neither enters nor leaves the system. Air is a relatively poor conductor of heat, so that when a large mass of air is lifted its drop in temperature is due only to the energy used in expanding. Any mass of 'dry' air lifted (by any means) to a higher level will cool at this rate. (*See fig. 23(a).*)

If the atmosphere has a higher lapse rate, any air which starts to move upwards will still be warmer than its surroundings and will continue to rise (as in fig. 23(b)) until it meets air which is warmer than itself.

On sunny days the surface temperature may rise quickly, heating the air in contact with it and forming a shallow layer which has an even higher **super adiabatic lapse rate.** This makes it buoyant, and once some of it is disturbed it starts to rise, forming a thermal. Because of the reduction of pressure with height it expands still further as it rises, cooling adiabatically. It will continue to rise until it reaches air at the same or higher temperature as itself, when it ceases to be buoyant.

If the air is saturated and cloud is forming, heat is released as the water vapour changes to water droplets. This extra heat released keeps the air a little warmer so that the lapse rate becomes less. This amount, known as the **wet** or **saturated adiabatic lapse**

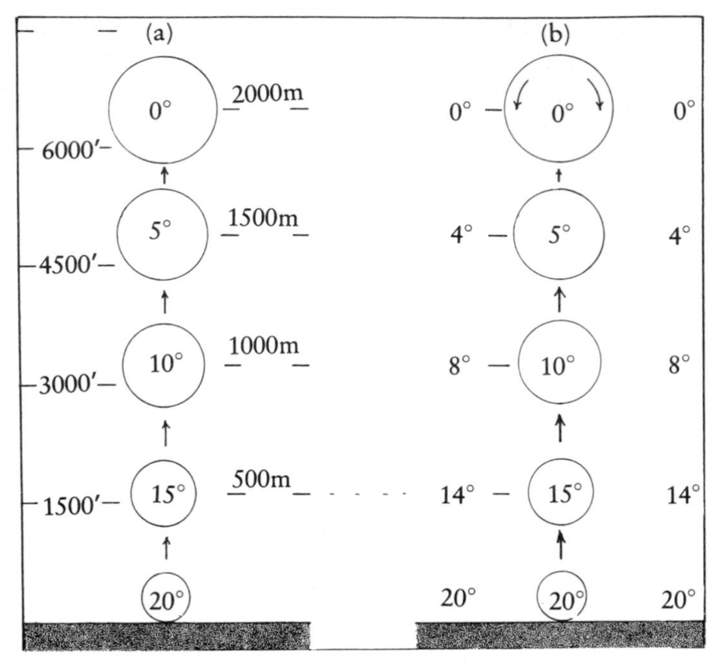

Fig 23 Stability and instability of the air. (a) The Dry Adiabatic Lapse Rate (DALR): 1°C drop in temperature for 100 metres increase in height. (b) A higher lapse rate makes the air unstable.

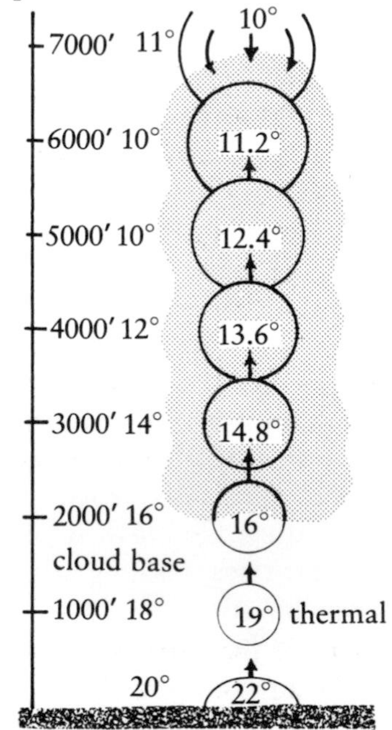

Fig 24 Conditional stability. With dry air, the thermal would have been limited to 2000 feet.

rate, is more variable and depends on the height and temperatures concerned. It is about 0.5°C per 100 metres (average in UK). This change in the lapse rate as the air becomes saturated can be important. The warmer the air, the more water vapour it can hold so that when warm moist air rises and becomes saturated, much larger amounts of heat are released than from cold air. This warm moist air can be stable to dry air, but becomes unstable if cloud forms so that the rising air remains at a higher temperature than its surroundings. (This is known as **conditional instability.**)

Fig. 24 gives an example of conditional stability. In this case, if the air had been dry, a thermal starting at 22 degrees could only rise to 2000 feet before being surrounded by air warmer than itself. If the air became saturated at 2000 feet the thermal could have reached over 6000 feet.

Height in feet	Measured temp. °C	Dry adiabatic	Wet adiabatic
8000	11.0		
7000	11.0		10.0
6000	10.0		11.2
5000	10.0		12.4
4000	12.0	10.0	13.6
3000	14.0	13.0	14.8
2000	16.0	16.0	condensation
1000	18.0	19.0	level
ground level	20.0	22.0	

Changes in the lapse rate and humidity largely determine the weather and are therefore of particular interest and importance to the forecaster. This information can be obtained by making 'ascents' in specially equipped aircraft or by sending up unmanned Radiosonde balloons. The results of these ascents are plotted on a special type of graph known as a Tephigram. This is used to predict the cloud base, cloud amounts, the surface temperatures required to start thermal activity and the strength of thermals likely on that day.

Inversions

Because of many factors, the state of the atmosphere is changing all the time so that the actual lapse rate is not a consistent figure. The temperature does not decrease consistently with height but varies considerably. With the advance of a warm front there will be a layer of much warmer air at a certain height. Also, after a clear night with the earth radiating off its heat and cooling rapidly, the layers of air close to the ground will be much cooler than normal, so that the temperature may actually increase with height during the first few thousand feet. This is known as a temperature **inversion** and it makes the air very stable, acting as a 'lid' to any thermals.

With an inversion, a thermal leaving the ground and rising will meet air which is warmer than itself. This will prevent it rising any further. If the inversion is only, say, at 700 feet, early thermal activity will be limited to that height. Further heating from the sun may raise the surface temperature until later thermals eventually reach that inversion warmer than the surrounding air and are therefore still buoyant. These thermals will break through the inversion and often quickly destroy it altogether, allowing later thermals to rise much higher. (*See fig. 25.*)

The early morning inversion after a cloudless night can be clearly seen as a layer of smoke haze above the horizon. If it is near the ground it may be easily broken down by the early thermal activity. Sometimes there are several inversions, each in turn acting as a barrier and limiting the height to which the thermals will

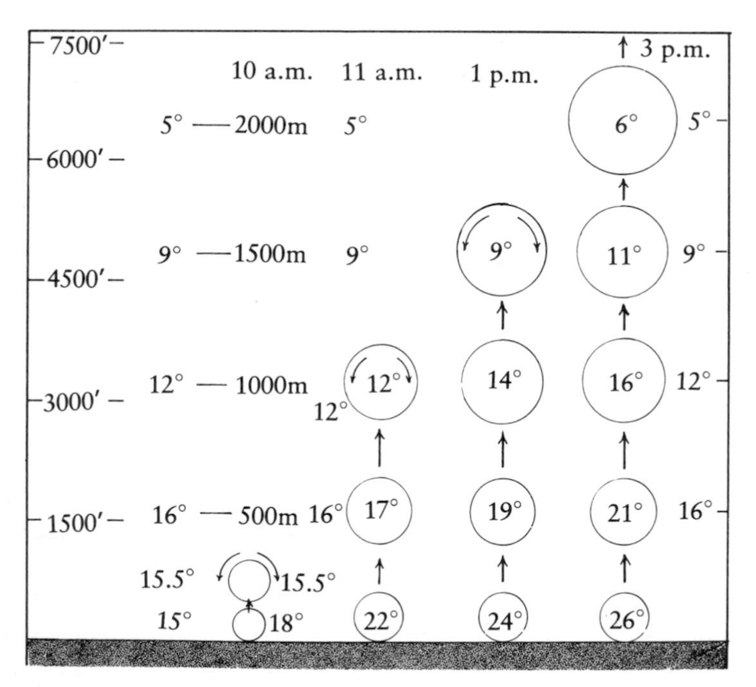

Fig 25 In this example a low level inversion limits thermals to less than 1000 feet at 10 am. By 11 am the low inversion has been broken. Later, as the surface heating increases, the thermals penetrate higher.

go. In this case the forecaster may predict with some confidence that, for example, thermals will reach 2500 feet when the surface temperature is 15°C at about 10.00Z, and 3500 feet when the surface temperature has increased to 21°C. Incidentally, most official forecasts use 'Z' or GMT (Greenwich Mean Time) so that to get local Summer Time in the UK, 1 hour must be added.

In the summer, a slow moving high pressure system becomes more and more stable with the inversions lowering until thermal activity is limited to less than a thousand feet, in spite of the very hot weather. After several days of hot unsoarable weather the heat wave is usually broken down by a violent thunderstorm or the approach of an active depression.

A layer of air with a constant temperature will have a similar effect of limiting or destroying thermals and is known as an **isothermal** layer.

Heavy showers and thunderstorms can develop whenever the air has a high lapse rate, making it very unstable to a great height and when there is sufficient moisture to produce a large cloud. The cumulonimbus clouds grow rapidly until they reach the tropopause, above which the temperature remains more or less constant. This is the edge of the stratosphere, at about 25–35,000 feet according to the season.

The formation of cloud results in a large release of heat energy. The latent heat which was absorbed, evaporating the moisture on the ground or from the sea, is released again, giving any thermal reaching cloud base an extra boost. This is the cause of the very powerful updrafts in showers and thunderstorms. Vast amounts of energy are involved. Just think of the weight of water falling during a cloud-burst, covering perhaps several square miles with an inch or so of rain. Thousands of tons of water have been evaporated and carried up to twenty or thirty thousand feet.

The forecaster obtains reports on the temperature and humidity from ground level up to great heights. These enable him to predict the height and amount of cloud at different levels and whether it is likely to spread out or produce storms.

Even with a Tephigram the interpretation of ascents is not a simple matter. So much depends on how quickly the surface temperature rises and on the way in which the airmass is modified during the day. Even a thin layer of upper cloud may delay the onset of thermal activity or prevent a layer of fog or low cloud from breaking up and dispersing as forecast.

For this reason a limited knowledge of meteorology cannot help very much in predicting soaring conditions. We just have to rely on our local forecaster to look at the ascents and make his predictions. Unfortunately in Great Britain there are only a few stations sending up these Sonde balloons so that the results may refer to the air over a hundred miles away from our flying area. By the time that the air has travelled over this distance it has often been modified sufficiently to upset the predictions of cloud amounts and heights.

5

CLOUDS

The formation of clouds

The amount of moisture that air can hold depends on its temperature, with warmer air able to hold more moisture than cold. When air is saturated with water vapour, any further cooling or addition of moisture will cause some to condense out in the form of cloud.

Dew point

Cloud forms whenever the air is cooled to the point where the amount of moisture in the air is sufficient to saturate it. The temperature to which a particular mass of air must be cooled for saturation to occur and for cloud to start to form is known as its **dew point**. If the temperature and dew point are almost equal, the air is very moist and fog or cloud will form if there is any further cooling or if more moisture is added by drizzle or rain.

Relative humidity

Sometimes the amount of moisture in the air is measured as **relative humidity.** The relative humidity is the ratio of the amount of water vapour in the air to the amount of water vapour required to saturate it at that temperature, expressed as a percentage. Relative humidity decreases with increasing temperature. If the relative humidity at ground level is nearly 100%, again the air is very close to being saturated and there would be a high probability of mist or fog forming unless the temperatures were rising.

The dew point and the relative humidity can be found from the readings of a Wet and Dry thermometer. This consists of two thermometers, one of which has its bulb covered in a muslin wick which is kept wet by a reservoir of distilled water. The evaporation around the bulb cools it according to the temperature and humidity of the air. Tables are used to determine the dew point and the relative humidity from the readings of the two thermometers.

Fog

There are various ways in which cooling in the atmosphere can take place, causing the air to saturate and cloud or fog to form. Fog forms if moist air moves over a cooler surface such as the sea (**advection** or **sea fog**) and at night in light winds if there are no clouds. Then the earth's heat is radiated out into space, cooling the ground and the air in contact with it (**radiation fog**).

When air is lifted it expands because of the reduction in pressure and this causes the drop in temperature. Since the cold air cannot hold so much moisture, the relative humidity rises until the air becomes saturated and cloud forms.

Air which is lifted by hills or a mountain range cools with height and if the air becomes saturated **orographic** cloud will form. Cloud on the surface formed like this is known as **hill fog.**

Thermal activity is caused by the uneven heating of the air near the ground. Air which becomes warmer than its surroundings will always tend to rise. This is because warm air expands and becomes less dense. As the air in a thermal rises, it expands and loses heat so that if there is sufficient moisture and the air is lifted far enough, the air in the thermal becomes saturated and cumulus type clouds are formed.

Types of cloud

Clouds are classified as high, medium or low according to the height of their base. (*See fig.* 26.) As pilots we should be concerned with all the cloud types, because some of them help to indicate the proximity of bad weather and inevitably cut off the sun's heating, spoiling any thermal activity.

High clouds (15 to 40,000 feet)

Clouds at high altitudes are largely composed of ice crystals and are known as **cirro** types:

Cirrus (*Ci*) is the high whispy cloud.

Cirrostratus(*Cs* or *Ci-St*) is the thin veil which causes a slight halo over the moon or sun but does not obscure them.

Cirrocumulus (*Cc* or *Ci-Cu*) is high cloud with a regular cellular pattern.

Medium height clouds (6500 to 23,000 feet)

These clouds all have a prefix **Alto:**

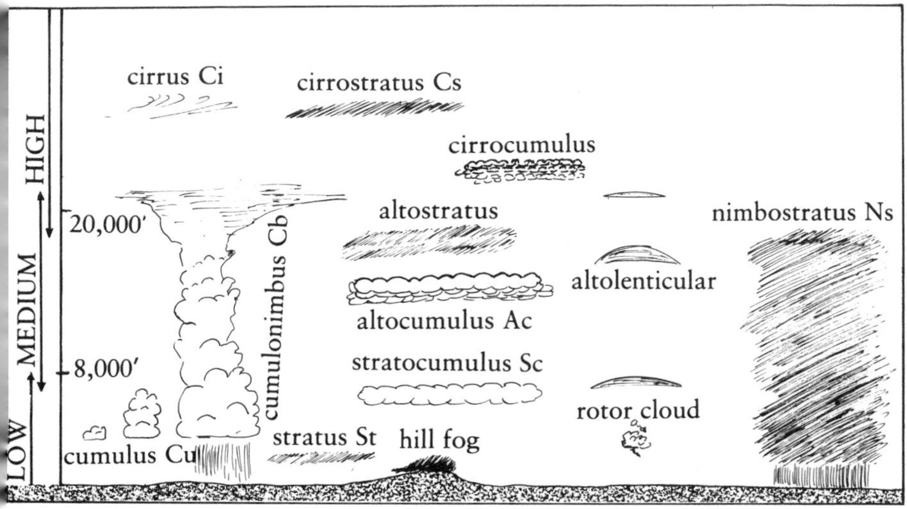

Fig 26 Types of cloud.

Altostratus (*As* or *Alto-St*) is an even-layer cloud at medium heights.

Altocumulus (*Ac* or *Alto-St*) is cloud forming at medium heights in a layer of regular cells rather than a smooth layer.

Stratocumulus (*Sc* or *St-Cu*) is a grey or whitish cloud layer consisting of rolls or cells.

Low clouds (0 to 8000 feet)

Stratus (*St*) is a layer of low cloud, often at hilltop level.

Cumulus (*Cu*) clouds are individual, fair weather clouds formed by thermal activity from the ground. The height of the cloud base varies from about 1000 feet to 6–8000 feet in the UK, but to over 20,000 feet in hotter countries and in desert regions.

Nimbus types are clouds with an extensive depth and usually with precipitation.

Cumulonimbus (*Cb*, but often referred to by pilots as **Cu-Nimb**) are heavy showers or thunderstorm clouds. These are large *Cu*, towering to great heights and having ice crystals in the upper areas forming an anvil shape.

Nimbostratus (*Ns*, *Nb-St* or *Nimb-St*) is very deep and extensive layer cloud completely blotting out the sun.

There are a number of sub-categories of cloud but it is most important to understand that cumulus types are ‘heap’ clouds formed when the air is lifted rapidly and locally. Layer clouds (the stratus family) are created when the lifting or the cooling of the air is gradual and widespread; **lenticular** clouds are lens shaped bars of cloud formed by the air being lifted in a wave system.

Cumulus cloud developing.

Two minutes later. (Photographs courtesy of John Simpson)

Spectacular lenticular clouds, at Portmoak. (Photograph courtesy of Alan Purnell)

Lenticular clouds at Minden, Nevada.

Cloud base

The cloud base or condensation level is the height at which the air is cooled down sufficiently for the moisture it contains to saturate it. After overnight cooling, which frequently results in dew, the air near the surface is relatively moist so that when it it lifted by flowing over hills or by thermal activity, cloud forms at low altitudes.

Typically, early on a summer morning the cloud base of the small cumulus formed by thermals will be only 1500 to 2000 feet. Even very weak thermals will form clouds at these heights and it is usually only when there are 'streets' of almost continuous cumulus clouds to help to organise the convection that the lift is strong enough for such low clouds to produce soarable conditions. However, the temperature rises and the ground dries out very rapidly; this reduces the relative humidity of the air so that the cloud base lifts quickly as the sun's heating increases, and the thermals become more organised and usable.

Cloud base can lower in a few minutes in showers.
(Photograph courtesy of Charles E. Brown)

By ten or eleven o'clock the base of the cumulus will have risen several thousand feet and the thermals will become stronger as the cloud base rises. If the temperature and dew point are known for a nearby airfield (from listening to VOLMET), the probable cloud base can be calculated. Multiply the difference in temperatures by 400 to obtain the cloud base in feet. For example, a difference of 10 degrees gives a cloud base of about 4000 feet.

Usually, the peak of conditions with the highest cloud base will occur fairly late in the afternoon and after the time for the maximum solar heating. Towards the end of the day the thermal activity fades gradually as the sun goes down, but the base of the decaying clouds remains high.

Any rain or shower will result in an increase in the relative humidity and a rapid lowering of the cloud base of the cumulus. Often a heavy shower or a period of rain will result in the air becoming saturated at very low altitudes and a layer of stratus cloud forming. If the air is already nearly saturated even slight rain or drizzle may result in a layer of low stratus forming in a few minutes, creating a very real hazard to any aircraft flying at the time, particularly in hilly country.

Variations

On some days the cloud base may vary considerably from place to place. Thermals starting from a very dry area will form shallow clouds with a much higher cloud base than those from ground which has had a recent shower. If there is a stable layer of air acting as a lid to prevent the cloud growing vertically, the cumulus clouds will tend to spread out and overconvect or overdevelop into a layer of altostratus or stratocumulus. This can cut off the heating from the sun so that the thermal activity weakens or dies out for a time until the cloud layer breaks up, and the whole process starts all over again after perhaps half an hour or so. This is known as **cycling** and to avoid being forced to land, the glider pilot must stay high and move quickly away from the areas in shadow towards an area which still has sunlight on the ground where the thermals are still active. Overconvecting is most pronounced when the inversion limiting the convection has moist air at its base before the convection begins. A very dry inversion will inhibit the spread-out of cloud and ensure that it does not persist.

In hot, dry areas, the more intense heating together with the drier air makes the cloud base much higher. Typically, the cloud base on a good day in Arizona or central Spain can be 15,000 to 20,000 feet. The cloud base of cumulus rarely exceeds 9,000 feet in the UK and averages 4–5,000 feet in summer.

It is worth noting that the spacing of thermal streets is generally much wider as the depth of the clouds increases. Glider pilots will know that although with a cloud base of 1500 feet the lift is barely good enough to maintain height, the thermals are often so close together that progress across country can be made with care. If the tops of the cumulus streets are 8–10,000 feet (3km.), the next street of lift is likely to be two to three times this height away (6–8km.) and although the lift is very strong, there are likely to be very strong downdraughts between these streets.

Showers

In order for showers to develop, the clouds have to build up to well above the freezing level. A low freezing level will therefore mean a high probability of showers occuring if the tops of any cloud are likely to go well above this level. For example, if the freezing level is at 10,000 feet and the tops of the cumulus clouds are forecast to be only 9000 feet, showers would be unlikely; whereas a similar forecast for the cloud with the freezing level at only 5000 feet will almost certainly mean a very wet day and very little flying, let alone soaring.

Icing

For the glider pilot, very severe icing will occur in large cumulus and cumulonimbus clouds. Cloud droplets do not freeze when carried above the freezing level unless they are disturbed or carried upwards to regions of very low temperature, e.g. below –20°C. Once above the freezing level the supercooled water droplets freeze on impact and ice builds up very quickly on all the leading edges of the aircraft.

With a modern glider it is extremely risky to get iced up by climbing in clouds above the freezing level. With the close tolerances and small gaps in the control surfaces even slight icing can jam them. Also, even a trace of ice will ruin the gliding performance, turning the gliding angle of a high performance model into that of a basic trainer. The effects of a lightning strike on a glass or carbon fibre glider are not known, and it may be unwise to tempt fate by flying near active storms.

In winter the showers may be of sleet or snow and the visibility can be reduced to fifty yards or less in a few moments by heavy snow. It is tempting to climb in the area of lift ahead of an advancing snow shower. This may be the only lift about on that day and, if it spreads quickly, you may get engulfed in it, forget which way is out and be in very real trouble in half a minute. Flying in snow

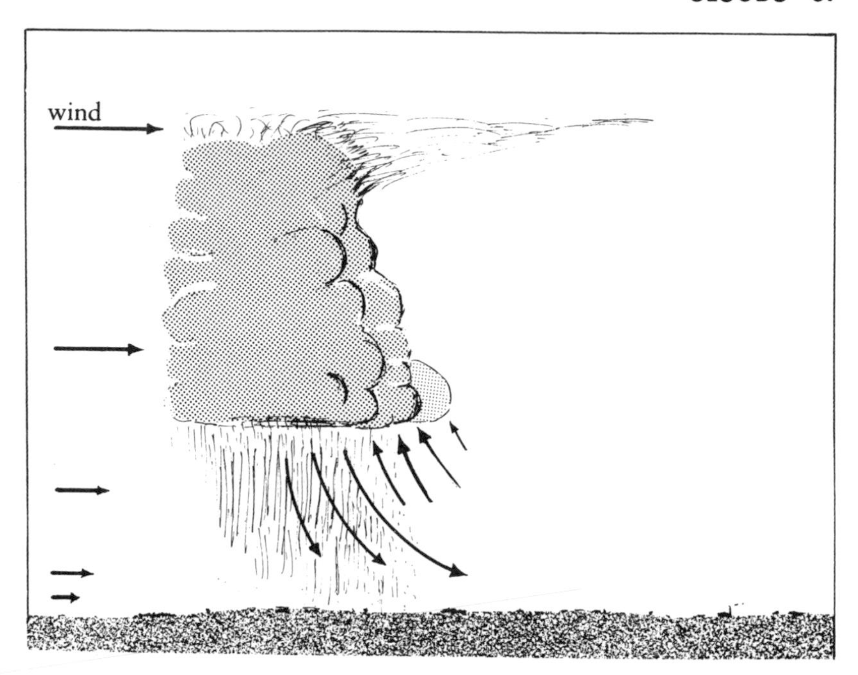

Fig 27 Strong rising air just ahead of a shower cloud.

is like flying in thick fog. There is no horizon and the ground becomes invisible. Think what it is like driving a car in heavy snow or torrential rain and use your imagination about getting caught flying in similar conditions. *Don't* get caught out.

Lift

If the showers are local and not too extensive, the falling rain or snow brings cool air down with it, moving along at the speed of the upper wind and spreading out as a thin layer of colder air. Because the wind at height is stronger than the surface wind, this cold air acts as a small cold front and produces strong lift just upwind of the edge of the rain. Using this kind of lift has its hazards. The lift is sometimes over a thousand feet per minute and so strong that you can be lifted up into the forming cloud even though the airbrakes are opened fully. If you take advantage of the lift over your home gliding site you will be drifted away down wind and cut off from the airfield by the curtain of heavy rain or snow. (*See fig. 27.*)

Landing

Sometimes it may be possible to work around the shower, but otherwise it may be wiser to make a run off downwind and select a good field for landing in case the winds are squally. Landing in the shower may involve very strong squally winds varying in direction from moment to moment. The cloud base may also lower very quickly and you may be lucky to make a safe landing in such adverse conditions with only a few hundred yards of visibility. Don't forget that when flying in rain or hail the performance of any type of aircraft will be degraded and the stalling speed will be considerably higher because of the wet wings. It is remarkably easy to be misled into believing that you have plenty of speed because of the extra noise of the rain or hail on the aircraft and to find that you are semi-stalled on the approach. Very heavy rain may also block the pitot or static sources, causing erroneous readings on the airspeed indicator and variometers.

Big showers can have a large area of sink in their wake. Although you might think that sinking air cannot possibly go down into the ground, some showers produce sink that hits the ground so strongly that trees are blown over in the airflow. Such **micro bursts**, as they are called, are common in the USA but can occur in the UK. They cause such extreme changes in the wind near the ground that a number of big jets have come to grief approaching to land.

In a glider it is always much safer to open the airbrakes and to get down onto the ground *before* the rain or snow reaches the edge of the airfield. Even then, assume a strong gusty wind, check the windsock and use plenty of extra speed for the approach. **Don't** try to spot-land, just get down in the middle of the field and make the best landing you can. Whatever you do, stay strapped in until the squall is past or help arrives.

Thunderstorms

In temperate climates, shower clouds which grow above 10,000 feet or so will frequently develop into active thunderstorms. Not only is there the risk of lightning strikes on the aircraft, but there is also the hazard of being damaged by hailstones. With the cable forming a perfect lightning conductor a thousand feet or more above the ground, any form of winching or car tow is extremely risky.

6

LOCAL TOPOGRAPHY

Sea breeze effects

During the summer, the land tends to warm up quickly, whereas the sea or large areas of water remain at much the same temperature. Thermal activity, particularly if the convection layer is deep, results in a slight reduction of the pressure over the land so that air is drawn in forming a sea breeze. In winter the sea temperature remains relatively warm so that even on a sunny day the contrast in temperature is not very marked and there is little or no sea breeze effect. At times cumulus streets can even form over the sea when very cold dry air moves out over the warmer sea and is warmed up by it.

Fronts

With an onshore wind, any sea breeze simply increases the strength of the wind, taking the cool, moist stable air further inland and spoiling the thermal activity. However, if the prevailing wind is parallel with the coast or has an offshore component, the sea air tends to move inland slowly during the day forming a **sea breeze front**. The cool air acts as a shallow cold front with rising air just ahead of it which can be used by gliders. (*See fig. 28.*)

At the edge of the front the cool moist air from the sea mixes with the drier air from inland, often forming distinctive hanging clouds with their base several thousand feet lower than the general cloud base. These clouds swirl visibly and mark the extreme edge of the lift. In dryer air, no cloud forms but often the front can be seen by the distinctive change in visibility. The stable sea air has a greatly reduced visibility and can appear as a layer of hazy, smoky air shaped like a hillside and advancing inland as a long line more or less parallel with the coast. If it can be seen, it is possible to 'hill soar' along the haze layer.

The exact position of the front can sometimes be detected by

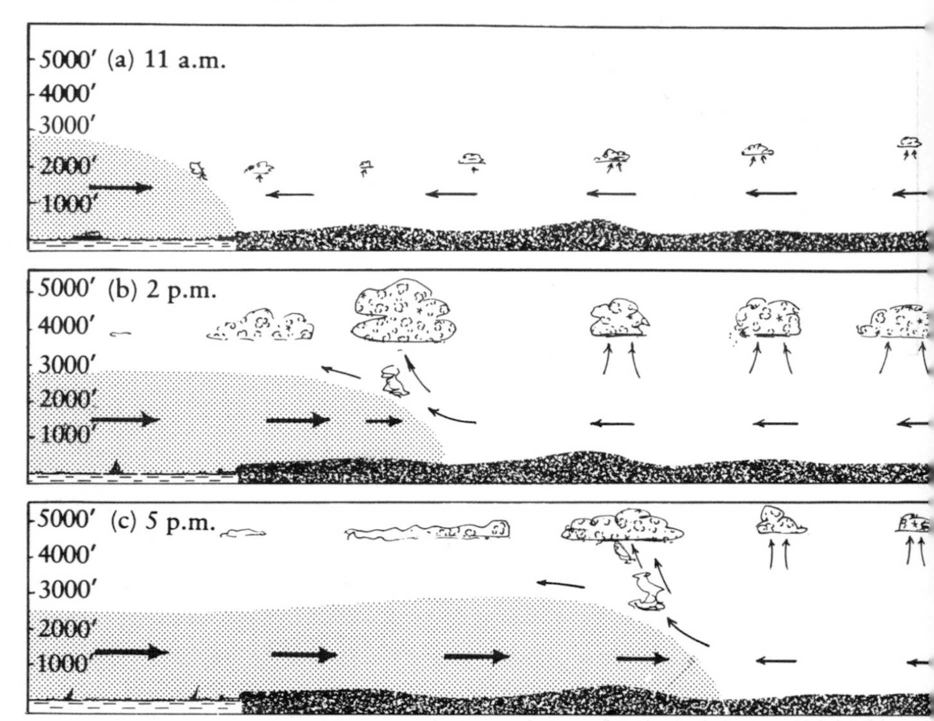

Fig 28 The development and movement of a sea breeze front. (a) In the morning some thermal activity up to the coastline. (b) Sea air begins to move inland, forming a front. (c) The front continues to move inland with 'curtain' cloud.

the wind direction. The complete change in the surface wind, which often occurs in a few minutes, may be seen from smoke in the vicinity. The wind in the sea air will be blowing in from the coast whereas the prevailing wind only a mile or so further inland on the other side of the front will still be in its original direction. In the sea air near the front the air is usually sinking rapidly and once a glider is in it, only a quick decision to increase speed and to fly further inland will prevent a landing before regaining the soarable conditions.

In the height of summer these sea breeze fronts penetrate 30 to 40 miles inland, gradually slowing down and then collapsing. Along a peninsula such as Devon and Cornwall and in northern England where the coastlines are not far apart, sea breezes can form on both coasts and move inland to meet and produce shower clouds. Similarly a collision may occur between the sea breeze front moving in from the south coast meeting a similar front moving in from the Thames estuary. This clash of two sea breeze

Fig 29 A sea breeze front seen from inland.

fronts is sometimes the cause of an unforecast thunderstorm in the London area.

In the cool moist sea air it is unusual for the sun's heating to be sufficient to produce worthwhile thermals. The warmer drier air above forms a very definite temperature inversion which acts as a lid and unless the heating is very intense, any thermals which do occur are weak and narrow. They are often broken and difficult for soaring and they do not usually go to more than two thousand feet or so before dispersing.

Curtain clouds hanging below the main base of the cumulus are a sign of entrainment of sea air into a convergence zone, and anywhere within 10 to 20 miles of a coastline they are a sure sign of a sea breeze front, a phenomenon that every British glider pilot must learn to recognise. (*See fig. 29.*)

Landing in the sea air has special hazards. The wind direction will be up to 180 degrees different from the general wind and, with sinking air and poor visibility in the sea air, it is only too easy to make the mistake of landing downwind with disastrous results.

Sea breeze fronts are mainly summer phenomena and seldom occur between October and April. If the prevailing offshore wind is strong, the front may form just off the coastline and remain there all day. In this case it may be possible to soar out over the sea in normal thermals or in the sea breeze front.

High ground effects

The wind direction and strength near hills can also be very variable. Whenever it can, the wind will blow round the ends of hills rather than over them. Landing in windy weather in hilly country is always risky because the wind is so unpredictable.

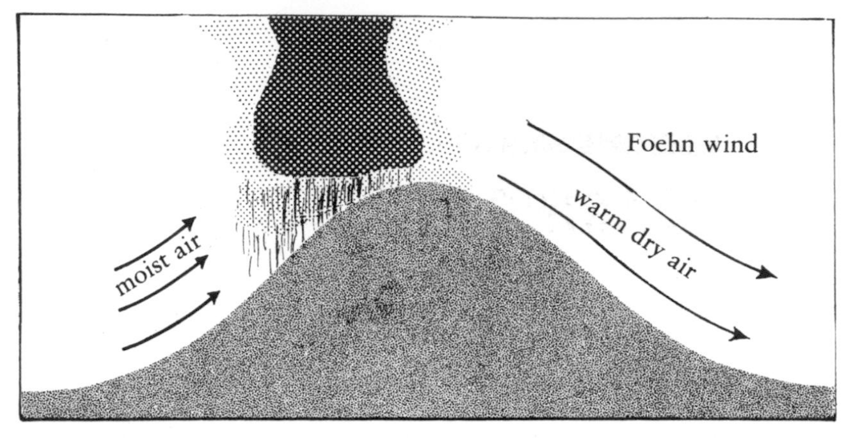

Fig 30 Foehn winds.

Air moving over high ground is lifted and cooled so that if it is moist, **orographic** cloud may form. If the lifting is gradual the cloud will be a layer type such as stratus or hill fog. Frequently, when flying at a hill site, cloud will suddenly form at hilltop level necessitating a very rapid descent if a safe landing is to be made. Any increase in the humidity of the air drifting over the hill may result in this cloud formation. At some hill sites cloud may start to form on the hills some distance upwind, giving a warning of potential trouble. However, in this situation the cloud does not drift towards the hill. Instead, the cloud forms around the hill top within a few seconds as the moist air arrives and is lifted over the hill.

Another effect caused by moist air flowing over high hills or mountains is illustrated in fig. 30. As the air is lifted, it cools, forming cloud and depositing much of its moisture as rain or snow. In addition, a lot of water can be caught by trees and other vegetation enveloped by the cloud. Having lost most of its moisture, the drier air descending in the lee of the high ground becomes much warmer than it was originally. (During the lifting it was saturated air and losing heat at the Saturated Adiabatic Lapse Rate of about 0.5°C per 100 metres; whereas when descending it is warming at the higher, Dry Adiabatic Lapse Rate of 1°C per 100 metres.) This is known as a FOEHN effect and is common in Scotland and in many places in the Alps and the Rockies.

7

SOARING CONDITIONS

Seasonal effects

The weather and the soaring conditions in Europe are particularly sensitive to the time of year. In winter, the sun's heating is seldom sufficient to produce usable thermals for more than an hour or so at midday. However, thermals do occur even with snow on the ground, but they are usually weak and infrequent. Instead of relying on thermal activity the soaring pilot must be content to fly in ridge and wave lift.

In the springtime there is often a period of 'spring easterlies' when unstable north easterly winds blow across the whole country producing surprisingly good soaring conditions with a high cloud base, shallow cumulus clouds and good streeting. Many years ago a record flight was made in these winds when a Skylark 2 flew from Lasham to Perranporth, flying the 190 miles in 3 hours and 40 minutes and only changing cloud streets twice in the whole flight. (By modern standards the Skylark is a very low performance glider.)

By Easter the sun's heating has become more intense and the conditions can be excellent. Many pilots miss some of the best days of the year by being out of practice and needing dual checks or by being caught with their aircraft unserviceable.

The very best conditions in England are on the day after the passage of a cold front with an approaching ridge of high pressure and northerly winds, giving consistent conditions over most of the country.

During the summer, after the passage of a cold front on the previous day, the thermal activity and cumulus formation may start as early as 8.00a.m. On other days the Cu will start to form about 10.00a.m. but will be too low to be workable for an hour or so.

By 11.00 or 12.00 the cloud base will be rising quickly. This means that if you really want to make a long distance glider flight you have to be ready to go before 9 o'clock to be sure of using the full potential of the day. Most of the basic preparations can be done beforehand. The barograph can be smoked ready to seal, the declaration can be partly filled in and the camera checked, all saving precious time in the morning. On many occasions you may have to wait several hours for the first cumulus to form but on the day of days you will be ready and can make the most of the possibilities.

However, many opportunities for interesting and often very fast cross-country flights are missed by assuming that the whole day has to be used. Frequently the conditions only get going during the afternoon after a cloudy start.

In the summer-time, if you are within 20–30 miles of a coastline, watch for signs of a sea breeze front. Long and fascinating soaring flights are possible along these fronts which often continue to work until late in the evening after all the other lift has ceased.

By June or July, the possibility of the conditions being spoilt by the ingress of the sea breeze has to be taken into account and flights close to any coastline become a doubtful proposition. For example, it is risky to choose a goal or turning point anywhere near the sea unless there is a fairly strong off-shore wind blowing. Later in the day it may be quite impossible to reach that area because the sea air has already moved further inland.

August and September may still produce very good soaring days but by then the days are shortening again, reducing the time available for very long flights. However, the farmers may be burning off their stubble and producing very strong, well marked thermals which are easy to locate and may compensate for the lack of daylight hours.

Thermals

On any sunny day some areas of ground are heated up a little more than others. High ground and particularly sun-facing slopes become very good thermal sources. The type and colour of the soil, the type of crop, the general drainage of the land and whether there has been a local shower recently are only a few of the factors which cause this uneven heating. The ground warms up the air in contact with it so that some areas of air become warmer than others and are therefore less dense. Any disturbance will encourage this air to break away from the surface and rise as a huge bubble, a series of bubbles or columns of rising air known as thermals.

Fig 31 Thermal bubbles. One type of thermal. Note that the internal motion is similar to a smoke ring.

In temperate climates it seems usual for thermals to be a series of individual bubbles of air and there is ample evidence to show that they soon develop an internal motion rather similar to a smoke ring. (*See fig. 31.*) The lift in the centre core is very much stronger than in the cap and is usually at least twice the rate of ascent of the whole bubble. This means that after a short while gliders using the thermal all arive at about the same level in the cap of the thermal, where they climb at much the same reduced rate. Other gliders joining the thermal at lower levels usually end up at the same height in a 'gaggle'.

On a good soaring day the thermal bubbles leave the same source fairly regularly so that the cumulus clouds form into a line running downwind. Recent research shows that even in quite a strong wind the air in the incipient thermal can act as an obstruction while it is in contact with the ground and that the winds may blow around it. For this reason, even in a wind the lift is often found directly above a good source at low altitude. However, once the bubble of buoyant air breaks away from the ground it drifts with the wind. Unless there is a marked shear in the wind with height, no correction should be needed to keep in the area of lift.

If the thermal rises high enough to form a cumulus cloud the best lift will usually be directly under the centre of the forming cloud (not always upwind of the cloud as is often assumed). As it rises, the warmer air in the thermal mixes with and entrains more of the surrounding air, diluting it and making it larger. If there is

Cumulus cloud forming over Didcot Power Station, an almost permanent thermal source. (Photograph courtesy of Hugh Hilditch)

a high lapse rate, the thermal will accelerate and keep on ascending until it meets a more stable layer. If it meets a layer with a constant temperature or with an inversion, the thermal will slow down and dissipate at that level.

When cloud forms, the release of latent heat gives the thermal an extra boost so that the cloud may continue to 'work' by a self-stoking process. In this case the cloud may develop, providing usable lift close to the cloud base, although no further thermal activity is supplying it from the ground. Of course, larger clouds may be being fed by a number of thermals simultaneously.

Cloud growth

If you observe the cumulus clouds on a summer day, you will notice that the first sign of a cloud forming is a patch of milky haze. This soon turns into a wispy patch of cloud developing upwards and growing denser. Unless the cloud is being fed almost continuously by new thermals, it begins to decay, evaporating at the sides and base, leaving the remnants of the cloud well above the normal cloud base. Often the cloud will be developing on one side and decaying on the other. Where the cloud is evaporating, a cold downdraft soon becomes established.

True cloud streets are not formed from individual hot spots. They need a definite lid to the convection and the winds increasing with height through the thermal layer. Then the convection becomes more organised and the cumulus clouds form regular streets lying directly up and down wind with areas of sink between the lines. Using these cloud streets it is sometimes possible to glide for 50 miles or more, cruising without circling and with-

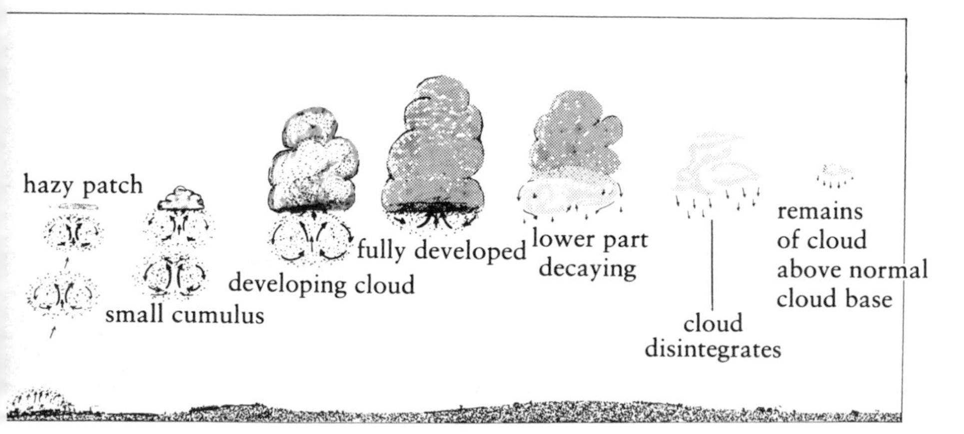

Fig 32 The life of a small cumulus cloud. Normally it takes a number of thermal bubbles to form even a small cloud.

out losing any height. Satellite pictures of cloud streets over the oceans show them to run for hundreds of miles but they have yet to be proven soarable.

Thermal sources

Woods and areas of water will generally warm up very slowly so that they are particularly poor sources of thermals. However, late in the evening when the open country is cooling off rapidly, the woods may begin to release their heat and can produce large areas of weak lift – the only lift at that time. In the tropics even lakes can produce thermals as they give off their heat at night. During the day, the upwind edges of lakes and other large areas of water often act as a good trigger for thermals, whereas the thermal activity downwind of them may be damped down for many miles. The cold dense air above the water acts as a barrier and triggers off any buoyant air nearby.

Sun-facing slopes will always heat up more quickly than flat land and if the slope faces south and is sheltered from the wind, it is likely to produce very strong thermals. Similarly any thermal coming from a source which has an extra form of heating will be stronger. Stubble fires, power stations, factories, towns and, of course, airfields, all tend to be extra good sources. However, the lift from these sources is seldom continuous. Usually they produce a burst of good lift every five or ten minutes followed by a lull with little or no lift.

The thermals may be just as strong or even stronger on days when no cloud forms. On these 'blue' days the heating is uninterrupted by cloud shadows and the only difficulty is guessing where the next thermal will be found. Again, unless the winds are very light and variable the lift will usually tend to be in streets lying up and down wind.

If there is a strong inversion forming a definite lid on the convection, any large cumulus clouds will tend to spread out at that level, cutting off the sun's heating for a while. As mentioned before, this over-convecting is more likely when the upper stable layer is rather moist.

Visibility

For forecasting purposes 'poor' visibility generally means a visibility of less than 1–2 miles. Poor visibility of over 1000 metres is defined as **mist** and less than 1000 metres as **fog**.

The visibility is usually exceptionally good after the passage of a cold front. The cooler unstable air results in visibilities of 30–40 miles. Visibility is reduced by any haze which may be the result of smoke pollution or high humidity. Strong inversions prevent the dust dispersing and, in the summer, anticyclonic conditions with persisting high pressure result in the haze becoming thicker and thicker.

Very poor visibility is also caused by the sea air behind the sea breeze front. This has a very high humidity, as well as collecting all the pollution into a comparatively shallow layer near the ground.

Any drizzle, rain or snow reduces the visibility to a few miles or less in a few minutes. Heavy rain or snow can reduce it to a few hundred yards making safe flying impossible without full instruments.

If the morning starts with hazy sunshine, the visibility will usually improve as the thermal activity breaks through the lower inversion and cumulus forms, spreading the dust and moisture up into the higher levels.

Upper cloud and even smoke or haze may cut off enough of the sun's heat to spoil the thermal activity. Areas downwind of large industrial cities will often have only weak thermals and should be crossed as quickly as possible or avoided altogether when the opportunity arises.

Even the strength of the upper winds can be important to the glider pilot because if the winds at high altitude are strong, patches of cirrus cloud may drift over quickly. With lighter winds even a small patch of upper cloud can cause a problem by cutting off a significant proportion of the sun's heating for a time.

Large cumulus and cumulonimbus clouds may cut off the direct heating but they also cause a general subsidence of the air over a vast area, often suppressing the thermal activity for 20–30 miles around the storm. Shower activity is difficult to predict and can easily put the best of glider pilots into a field. A recent shower will stabilise the air over the wetted area and it is almost impossible to tell where and when a shower has occurred. Usually showers tend to favour high ground and develop late in the day as the temperatures get higher.

When showers are forecast it is normally best to take off as early as possible and to try to complete the flight before the showers develop and become troublesome. There is always a certain element of luck when flying a glider across country on a showery day. Arriving a few minutes early at a turning point may result in having a superb climb and an easy flight home, whereas minutes later the rain may be falling and result in it being impossible even to stay up.

Wave conditions

Lee waves

Lee waves formed by hills and mountains may occur at any time of the year, but are usually more common in the autumn and spring when the winds are stronger and the air is a little more stable. In general, the stronger the wind, the longer the wave length with a typical distance between wave crests of between 3 and 10 miles.

It is usual to have to fly more or less into the wind to stay in the up current of the wave and to have to fly much faster at height to avoid drifting backwards into the sink. So, in many respects, this is similar to hill soaring except that if there are no clouds, the 'hill' itself is invisible. Unless the wave is very steep, circling in the lift results in drifting quickly into the down area.

The airflow in both the up and down areas of the waves is usually incredibly smooth, but there can be rough air at the edge of the wave flow or if the wave steepens and breaks. It is not unusual to find lift of 600 to 800 feet per minute if the wind is strong, and rates of climb of up to 2000 feet per minute have been observed. This means an even higher rate of descent in the down parts of the waves and accounts for some of the accidents which occur in hilly country.

In ideal conditions the wave lift can continue to over 30,000 feet

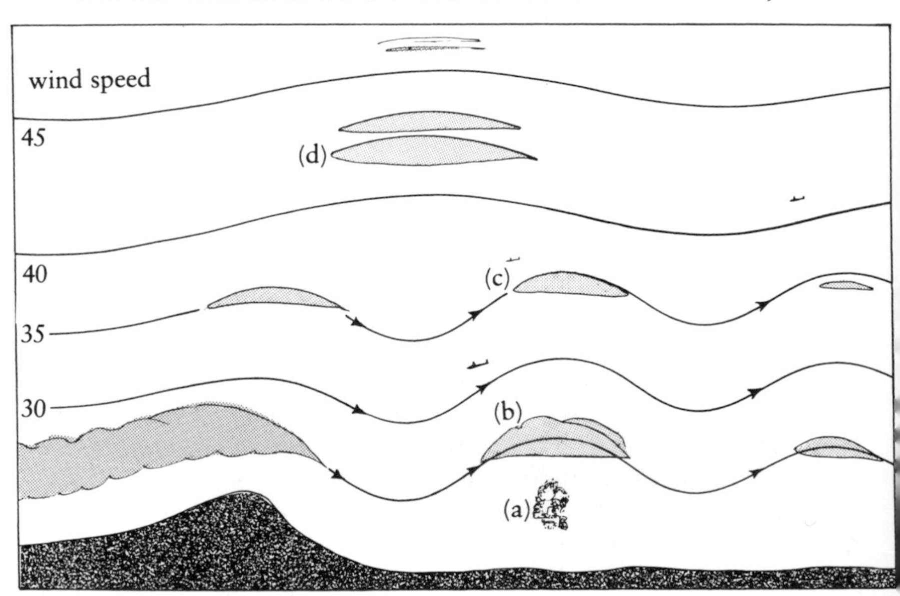

Fig 33 Lee waves. (a) Rotor cloud. (b) Low level lenticulars. (c) Medium level alto lenticulars. (d) High level cirrus lenticulars.

in the lee of quite small ridges and mountains. Generally, if there is usable wave lift in one part of the country it will be occurring over a wide area and in the lee of many of the hills. Often it is possible in a glider to jump from one system to another, covering large distances at very high average speeds. Cruising at high altitudes gives a bonus in true airspeeds because of the lower density of the air at height. At 20,000 feet, for example, flying at an indicated speed of 60 knots, the true airspeed is over 80 knots and at 40,000 feet the true airspeed would be almost 120 knots. When these speeds are added to a wind speed of 30–40 knots or so, the potential for long distance glider flights can be realised. Caution is needed flying gliders at very high altitudes to restrict the indicated airspeed, to avoid excessively high true airspeeds and the risks of 'flutter'. (Modern gliders are usually placarded to restrict speeds above 15,000 feet, to limit the true airspeed to below 130 knots.) It should be noted that the indicated stalling speed for any aircraft remains the same at all heights, but that flutter depends on the true airspeed and not what the instrument reads at height.

If there is sufficient moisture in the air, cloud will form as the air is lifted in the waves and will evaporate as it descends and warms up again. Usually the tops of these **lenticular** clouds have a unique smooth appearance and a distinctive lens shape when seen from a distance. (*See fig. 33.*) If the cloud is an extensive layer, a stationary clear slot or slots may be the only indication of the wave system visible from the ground. These clouds are recognisable because they form in stationary bars in the lee of the ridge of hills setting off the waves. Sometimes they will appear similar to a line of cumulus lying more or less across the wind, but careful observation shows that in spite of the wind, the bars or lines of cumulus remain virtually stationary relative to the ground. The gaps between these clouds will vary as the humidity of the air changes. If a very moist air mass arrives the gaps will close in a few minutes, trapping the unwary pilot above solid cloud and frequently over hilly country.

Although lee waves can be forecast, they often occur at other times when theoretically they are unlikely. So, if the forecaster says that waves are unlikely and the sky is full of lenticular clouds — believe what you see: it is worth looking for wave lift. However, good-looking lenticular clouds can be deceptive as they are often formed by weak waves which may prove to be unsoarable. Also, they may only last a few minutes and may disappear completely at any moment, if the conditions change so that the wave system collapses. The best lift is often found near scrappy-looking clouds which at first sight do not even look like wave clouds.

The requirement for the formation of these waves is mainly that the winds need to be in a fairly constant direction and increasing with height. The strongest waves will occur when the wind direction is at right angles to the line of the hills. A shallow unstable layer at low level, with a more stable layer above it, is most likely to promote wave formation. The best low level waves are often when the top of the stable air is at about the level of the highest hill tops. These conditions are common in a warm sector and also when the winds are strong enough, on the edge of an anticyclone.

Standing waves

The distinctive lenticular clouds may form at any height where the air is moist, and they lie in bars parallel to the ridge of hills or mountains forming the wave and not necessarily at right angles to the wind. In spite of the strong winds, these clouds remain almost stationary and do not drift with the wind. For this reason they are often known as **standing** waves. Wave lift exists below and above these clouds but also occurs in clear air if the air is dry. Note that the best lift is found just upwind of the edge of the lenticular clouds, about a quarter of the wave length ahead of the middle of the cloud. (*See fig. 34.*) The strength of the up and down draughts in a wave system is about the same with the strength decreasing with each successive wave downwind, unless they are boosted by being in phase with a further range of hills. However, in terms of rates of climb and descent, the glider pilot will find the sink rate about double the climb rate unless the wave is very strong. For example, with the air moving up and down at 500 feet per minute the average glider will climb at about $500 - 150 = 350$ feet per minute, but will sink at $500 + 150 = 650$ feet per minute.

For the aeroplane pilot it is important to realise that the sinking air may often be strong enough to prevent any climb, even at full power. Caught in the down of a wave, it is important to turn downwind to find better conditions and not to fly parallel to the ridge causing the wave system unless an area of rising air is found.

Wave systems

Wave systems can be complicated, with waves of a different length superimposed above the lower waves. Moving up into the upper system may involve climbing as high as possible before driving forward hoping to find the next area of lift. Tactics for both hill and wave soaring are different from thermalling, where it pays to be selective and only use the best of the lift. In wave soaring it usually pays to be patient and gain as much height as

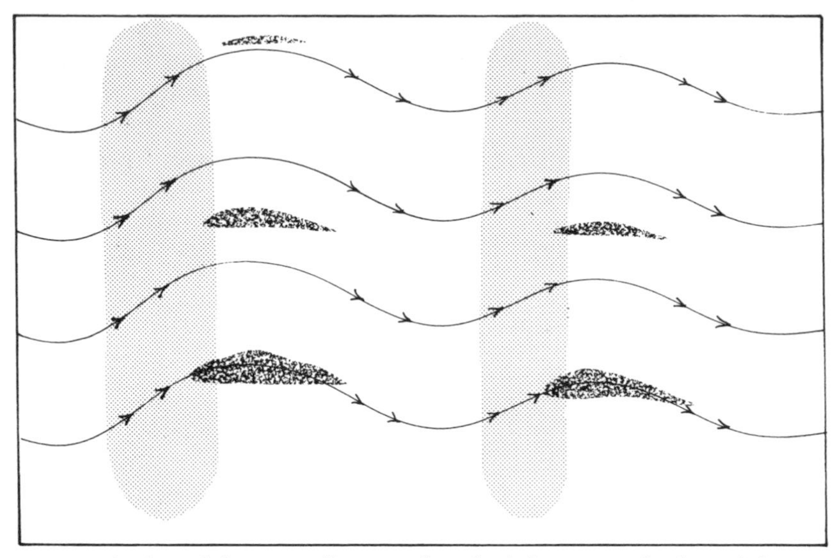

Fig 34 The best lift is usually just ahead of the upwind edge of the lenticular clouds.

possible before attempting to move on in search of anything better. Moving upwind to the next area of lift through the downdraft may cost many thousands of feet. Very often the waves slope forwards, making it necessary to move forwards to stay in the best lift as more height is gained.

At low levels, the airflow can be violently turbulent below the main wave system and in particular just below the primary wave in the lee of the hill. In strong winds, the **rotor** flow as it is called, may be dangerous for light aircraft and even for the strongest gliders, and it must always be treated with respect. If the air is moist enough, the worst area may be marked by a very ragged-looking cloud which at first sight may appear to be a normal cumulus. The violent swirling of the air can usually be seen in the rotor cloud, if one forms. Less turbulent conditions can usually be found if it is possible to fly along to one end of the ridge or at the end of the lenticular or rotor clouds where the lift, sink, and rotor will be weaker.

In normal thermal conditions, if cumulus cloud streets start to form across the prevailing wind there is almost certainly some wave activity. This may have been triggered off by hills or mountains fifty or sixty miles upwind. While not being strong enough to be usable, weak waves often kill the lift under even good-looking cumulus, making soaring extremely difficult. Sometimes the waves are stronger and can be used by climbing up below the

Lenticulars and rotor clouds in the wave system above Minden, Nevada.

Gliding in a Super Cub, with the propellor stopped, the author gained over 7,000 feet with a rate of climb of almost 600 feet per minute.

larger cumulus and then working upwind and into the wave. If cloud flying is permissible, climb up into the cloud and fly out up wind to try to get into the wave.

Waves can also occur over flat country and may not be associated with hills or mountains. (*See fig. 35.*)

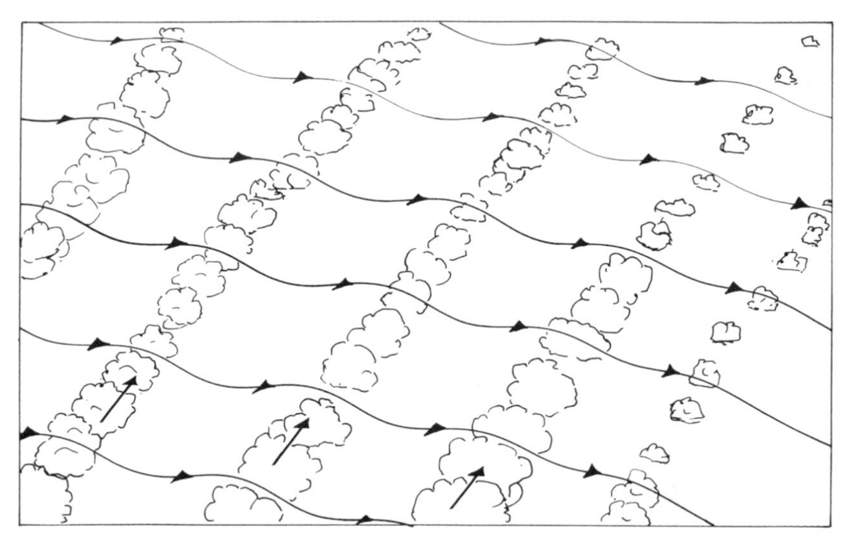

Fig 35 Wave formation above cumulus cloud streets. Here the waves are caused by the large change in wind direction just above the cloud streets which are forming up and down the lower wind.

They may form above normal cumulus streets if there is a marked change in wind direction just above the cloud tops. In this case the clouds themselves act as a barrier to set up the wave motion. It is possible in these conditions for a glider to climb up the face of the clouds on the upwind side and to continue climbing above the clouds to 10–15,000 feet. With normal cumulus the lift may continue in the cloud but not outside it. There have now been many examples of soarable waves above cumulus streets when the wind direction was parallel to the streets and increasing with height. If you climb into the cumulus at the upwind end of a street you may find the wave lift on the upwind face of the cloud.

Ridge lift and hill soaring

For good hill soaring, almost any day with a fairly strong and consistent wind blowing up the slope of a ridge of hills will give soarable conditions. The position of the best lift varies according to the exact wind direction and strength, the hill size and shape and the stability of the airflow, but is usually above and in front of the hill as shown in Fig. 36. A long unbroken ridge of hills is far more effective than an isolated hill or short ridge.

It is normal to have an area of severe turbulence and down draught behind the crest of the hill and this may extend to a mile or more downwind. If the top of the ridge is sharp, there may even

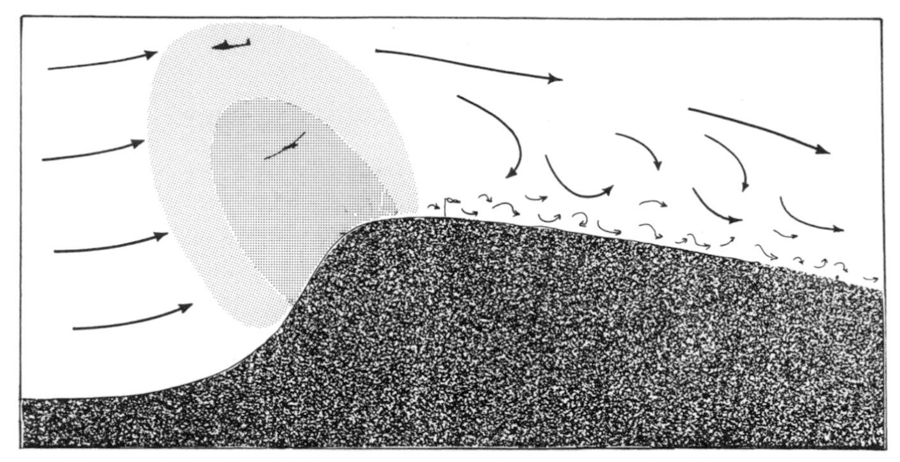

Fig 36 Hill lift. This diagram shows the best area for lift and the turbulent curl-over behind the hill top.

be a region just behind the top where the wind direction is reversed close to the ground, making landings dangerous. Landing at the top of a ridge in strong winds is always critical, with the very strong wind gradient and turbulence causing problems.

Strong thermal activity will usually make the hill life inconsistent and unreliable. Every time that a thermal or a good cumulus cloud drifts over the hill the lift will intensify for a few minutes. But between thermals the hill lift will deteriorate and only moving quickly along the hill to a better area will prevent a serious loss of height.

Wave activity from other hills upwind affects the normal hill lift. A slight change in wave length in a wave system set up by hills many miles upwind may put your hill out of phase with the wave. Without warning the down current from the wave may kill the hill lift in spite of a strong wind. This can happen in a few seconds and it results in many unpremeditated landings in gliders at some hill sites. Even a few minutes later the wave pattern may change slightly and unusually good soaring conditions can occur as the wave comes back into phase with the hill lift. Certainly, if you are hill soaring and are able to climb to more than about twice the height of the hill, it is most likely that you are soaring in thermal or wave. If it is very smooth air, be patient and work up as much height as possible before attempting to move forwards to reach the wave. If it is thermal, move out away from the hillside in a series of S-turns in the lift until it is clear to use it by circling normally.

8

ADVICE ON USING SOARING CONDITIONS

Thermal activity exists on almost every summer day and the energy available can be used by gliders, light aircraft of all kinds and even parachutes. Very few power pilots ever use thermal activity to their advantage, but it can be done and is not difficult.

In powered aircraft

Most aircraft have some kind of vertical speed indicator (VSI), variometer or sensitive altimeter which can be used to confirm that the aircraft is in rising air ('lift' to the glider pilot). It is then a matter of knowing how to use it for gaining height or reducing the need for so much power.

Using lift

Generally, using thermal lift effectively means slowing down to minimum flying speed and circling tightly. Although the areas of lift may be quite extensive, the really useful cores are usually only a few hundred feet across so that careful centring is needed to get the best advantage. At higher speeds the radius of turn is too large to keep the aircraft inside the area of lift and in straight flight a slow speed is desirable to keep the aircraft in the area for as long as possible.

The ordinary aeroplane VSI has far too much lag to be of much help except to confirm the overall results. At cruising speeds of more than about 80 knots, rising air is only felt as a jolt. Flying at 50–60 knots, however, with a little practice most pilots will be able to detect the edge of the lift by the extra 'g' as the aircraft accelerates upwards. This is the moment to start a steep turn at the minimum safe flying speed to try to keep inside the limited area of rising air. (*See fig. 37.*) Too much speed will result in a large radius turn and the probability of flying out of the lift and into the sinking air alongside.

Stubble fires provide very strong lift and usually form active cumulus clouds. (Photographs courtesy of Hugh Hilditch)

Unlike glider variometers, the vertical speed indicators in powered aircraft are not total energy compensated. This means that any pitching movements in the turns will result in very misleading indications on the instrument. As the nose is pulled up the VSI will soon show a vastly improved rate of climb for a few seconds as the

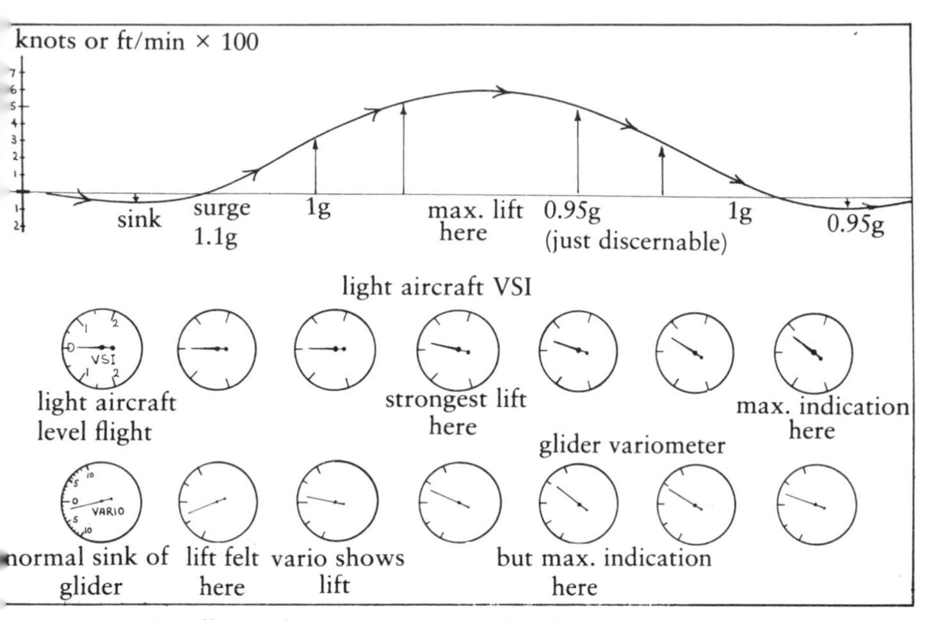

Fig 37 The effects of lag in the VSI and a glider variometer. Never wait for the best reading before starting to turn to use the lift.

speed is lost. A few seconds later, as the nose is lowered to regain speed, the VSI will show little or no climb at all. So a steady turn is essential if the true position of any lift is to be determined. It is only the average indication which can tell the pilot whether the circling is worthwhile.

Usually the lift is strongest in a core a few hundred feet across and the machine with the lowest circling speed will be able to use this area to the best advantage. For example, a hang glider may be able to outclimb a far more efficient sailplane by its ability to circle at a much lower speed in far smaller circles. Similarly, by using thermal lift a microlight normally climbing at only a few hundred feet per minute will often outclimb a light aircraft such as a Cessna. The loss in efficiency caused by turning steeply is almost always offset by being able to get into the stronger lift, and 35 to 40 degrees of bank is the minimum to try at first. In many machines lowering a few degrees of flap will allow a slower flying speed giving a smaller radius of turn.

Very small adjustments to the turns can be used to centre into the best areas, but remember that the VSI is indicating what went on some time before and is little help in determining which way to move. Use your sense of feel to recognise the best places and then on the next circle move out slightly towards the promising area. At most, bring the wings level for the briefest moment using full aileron control to roll out of and back into the turn, to move towards the better lift. Straightening up for a few seconds will usually take you right out of the best area.

A clever demonstration pilot can often give a rather under-powered machine an impressive rate of climb by using a thermal. Most competent towplane pilots use thermals to improve their climb by circling while towing the gliders, even though they are forced to fly at 50 to 60 knots which gives rather a large turning radius. It should be noted, however, that circling continously to use thermals may not be popular with any passengers. The air in and around the best thermals is usually the most turbulent, whereas the air away from the cumulus clouds in the blue is far smoother, and therefore it is kinder to avoid thermalling with inexperienced flyers. However, deliberately flying along under a cloud street can also give a vast improvement in climb or a higher cruising speed with less throttle.

On a day with well formed cumulus clouds it should be easy to find good lift by flying slowly below them. Not every cloud will work, but in a powered machine it is easy to move on from cloud to cloud until good lift is found. Usually the lift can be used to boost the climb from about 6–800 feet up to the cloud base, although it may be necessary to re-centre from time to time during a long climb. In some types of light aircraft it may be necessary to reduce power to prevent the engine from overheating. Even so it will pay to use the lift to climb rapidly unless you are in a hurry to go somewhere. Typically, with practice, you can add an extra 500 feet per minute to your rate of climb by using thermals and if you are just flying locally, it makes the flying more interesting than just flying along straight and level.

Similarly, it can be fun and well worthwhile to divert and cruise along a long ridge of hills using the hill lift, or in the lee of the hills if there is wave lift. The lower the normal rate of climb of the aircraft you are flying, the more there is to be gained by using this free energy.

In gliders and low speed aircraft

Although the expert soaring pilot seems always able to find and make good use of the soaring conditions, inexperienced glider pilots often fail to stay up on a good day. In most cases the reasons for their failures are as follows:

1 not initiating the turn early enough;
2 not turning steeply enough;

3 flying too fast in the turns;
4 poor searching techniques.

1 Not initiating the turn early enough

The first indication of flying into lift is usually a slight feeling of turbulence followed by an upward surge and an increase in 'g'. Because of the time lag in the variometer, the change in the rate of descent will take a few seconds to show on the instrument. Most beginners wait far too long before starting to turn, even waiting until the instrument shows the maximum rate of climb before applying the bank. By this time the glider is almost always through the lift and into the adjoining sink. **Never** wait for the best reading unless you are prepared to lose that area of lift and are high enough to move on to another area. (*See fig. 37 on page 79.*)

It is usually best to start turning as soon as the variometer reading has confirmed your sense of feel by moving up towards zero, and to make a well banked turn immediately. This will give you a very good chance of being close to the lift you detected. Then, once you have decided in which direction you need to move to get centred, straighten up by bringing the wings level for a fraction of a second before resuming the turn. A series of small corrections is usually needed.

Do not watch the variometer while making a centring movement. As the wings are brought level the glider is more efficient and a good variometer system will always show an extra rate of climb or less sink as this is done. Glance at the instrument to get readings at four positions of the circle to determine where the lift lies, but remember that you want to move the *whole* circle towards the best area of lift, which means straightening up more than 90 degrees before the area indicated by the variometer. (*See fig. 38.*)

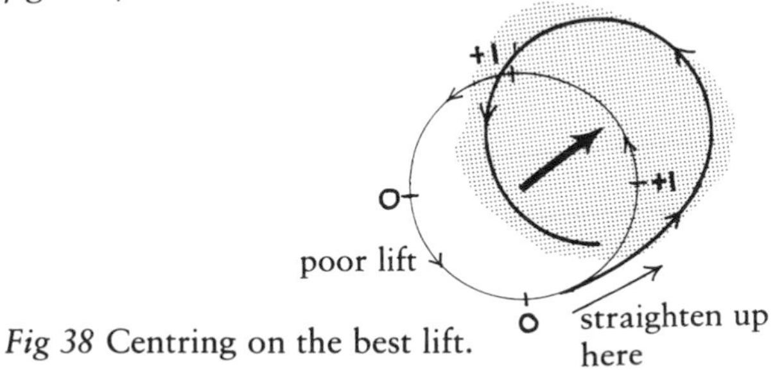

Fig 38 Centring on the best lift.

Fig. 39 shows that the first turn does not bring you back into the area where you first detected lift. Very often as you make the first turn, the lift immediately turns to sink. In this case keep the turn

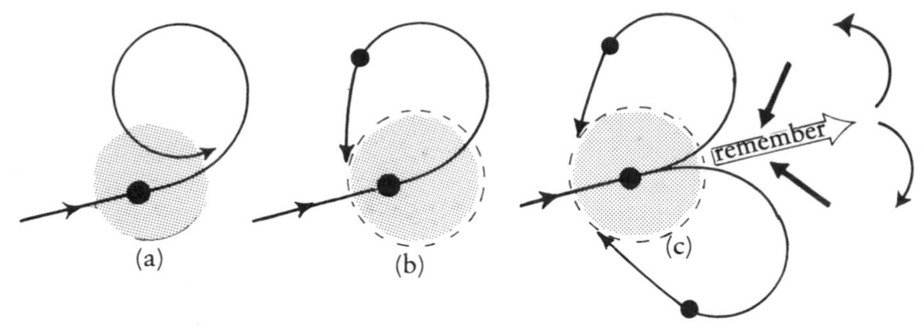

Fig 39 Moving back to find the lift. (a) The initial turn takes you out to one side of the lift. (b) Straighten up after turning a little less than 270 degrees and then continue the turn. (c) Remember the original direction and be ready to correct if the lift turns to sink.

going and straighten up momentarily a little before you have flown round 270 degrees. This should bring you back to the original lift. This is easy to do if you remember to notice in which direction you were flying as you flew into the rising air. **Never** reverse the turn as this will almost always result in losing the area of lift altogether.

2 Not turning steeply enough

The radius of turn of any aircraft depends only on the flying speed and the angle of bank. In order to stay inside the area of lift it is best to make the turns at the minimum speed necessary to have good control. In most gliders try to keep just above the speed at which the pre-stall buffet begins. Apply the bank quickly and do not use less than 30 degrees of bank. Use more if you can still turn steadily. Obviously if you turn quickly and steeply you will keep closer to the bit of lift you found and you are more likely to stay inside it. If you are not used to circling at such low speeds you may find the high rate of turn misleading. At 50 knots a 30 degree bank will result in a 360 degree turn every 28 seconds. This would be a rather gentle turn for thermalling but you might be misled into believing that you are turning steeply enough by the rate of turn. (This is four times the normal powered aircraft rate 1 turn.)

Fig. 40 shows two good ways of centring on the best lift. If at anytime you are turning at an angle of bank which allows you to steepen it by a significant amount, just steepen the turn as you feel a surge of better lift. Otherwise remember where the best lift was felt and open out the turn at least 90 degrees before you reach that position on the next circle.

It is important to realise that to get the best results involves

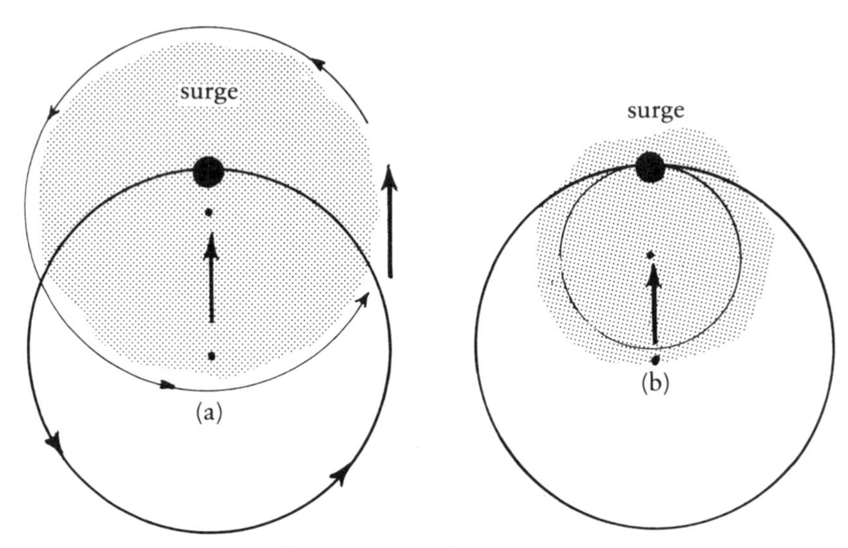

Fig 40 Two methods of centring on the lift (a) Straighten up momentarily at least 90 degrees before the position you felt the best lift. (b) Tighten the turn when you feel a surge of stronger lift.

finding and using the strong, narrow cores of lift. Although the rate of sink always increases with an increase in the angle of bank, this is insignificant compared with the improvement in the rate of climb resulting from being able to stay in the stronger lift. Getting a high rate of climb is almost always a matter of making tight turns and not a matter of turning efficiently. Do not make any large corrections to adjust your position once you are gaining height. One large correction in the wrong direction and you have lost the lift altogether with little hope of finding it again.

3 Flying too fast in the turns

Flying with excess speed increases the radius of the turns and also the rate of descent. Fluctuating speeds result in circles which are not concentric and in losing the lift altogether. If you cannot turn steeply, accurately and smoothly, use a more gentle angle of bank so that you can keep a steady turn. Because of the special characteristics of high aspect ratio wings (long wing span to chord ratio), in gliders it is safe to circle steeply within a few knots of the pre-stall buffet speed. Having established the turn it is best to slow down until the start of the buffet is felt and then to fly just above that speed. This is not safe in the average light or microlight aircraft as any turbulence or slight inaccuracy could result in an incipient spin. For this reason in these types of aircraft it would be most unwise for an inexperienced pilot to circle very slowly below 1000 feet or so to use thermals.

Whereas it is the correct thing to increase speed when flying through bad, sinking air and to slow down in lift, this only applies when flying straight. **Never** increase the speed in sink while you are turning as this will take you even further away from the lift and into the sink.

If you are downwind of the field and anxious about reaching it against the wind, the speed should be increased by about half the wind speed. This will give you the best distance against a strong wind. But while you are within easy reach of your field, do **not** increase speed flying into wind. This only results in having a further loss of height which has to be regained.

4 Poor searching techniques

If you are on a gliding site, watch the experts and note how they fly purposefully towards a definite cloud and find lift almost every time. After your launch look for other gliders circling nearby. They are in lift so don't ignore them and think you will find something of your own. Join them by searching below and circling in the same direction. If you don't start to climb or at least hold your height, open out into a wider turn to find better lift. If this doesn't work, move on to another area without delay. Continuing to circle losing height is a quick way down to a landing. If there are no other gliders marking a thermal, look up at the nearest clouds and fly immediately below the middle of the nearest well formed cloud you can reach. The best clouds are usually the ones with a darker base.

If there are no clouds to search under, cruise in straight lines searching in long zig-zags across wind. Do not fly too fast as this makes detection of a thermal more difficult and you will probably fly right through one before you have noticed it and started to turn. No racing stuff here; cruise at about the best glide angle speed, 45–50 knots for most modern machines.

Watch carefully for any tendency for the glider to be tipped into a turn indicating stronger rising air under that wingtip. Always turn against the tipping motion immediately as this will usually take you into the lift.

Keep referring back to your landing ground and do not fly away from it for more than a few seconds without checking your position and angle to it. Make your last search within easy gliding reach of your field and do not attempt to use any lift you find once you have made the decision to land.

Additional hazards for glider pilots converting to powered aircraft

In some ways the pure glider pilot is protected from many of the hazards of flying into dangerous conditions because of the need to find lift to stay up. Apart from hill and wave lift, the moment the sun's heating is cut off by any continuous layer of cloud, a landing is almost inevitable. Although training flights by winch or car launching often continue in relatively low cloud and poor visibility, these are five-minute, 'up, round and down' flights so that there is little risk of being caught in a sudden deteriorating situation. The majority of glider pilots have little or no experience of the problems of flying in bad weather. There is therefore a serious risk that when they fly powered aircraft they may not recognise dangerous conditions until it is too late.

Weather hazards

For example, the approach of a warm front is very gradual if viewed from the ground, with the lowering cloud base and start of the rain belt taking several hours to move in at 15 knots or so. This process is speeded up four times or more if you are flying towards it in a light aircraft at a hundred miles an hour and within minutes the conditions can change from being able to see ahead some 5 to 10 miles, to a 500 foot cloud base covering any low hills, and such bad visibility that it makes it very difficult even to pick a field for a safe landing. It can never be assumed that the deteriorating conditions can be avoided by turning round and running back out of them. Rain areas and showers often develop so quickly that the weather behind has become just as bad as that ahead.

With a powered aircraft a forecast **must** be obtained before making anything other than a local flight in fine weather. Low cloud and hill fog will form first over higher ground and can spread into an unbroken layer. In particular, mist and fog can form very quickly in the evenings as the air cools down and it may be your own airfield that is fogged in when you come back to it after a half-hour's 'jolly'.

It is always very risky to fly into worsening weather conditions because of the possibility of the timing of the forecast being inaccurate. Fronts have a habit of moving erratically, and instead of arriving at your destination just ahead of the bad conditions it is not unusual to have to divert because the front has arrived early. Cloud flying in a glider when the cloud base is three or four thousand feet is a very different matter from flying into cloud at a few hundred feet with icing conditions and hilly country ahead.

Hazy conditions can also be a problem at the higher speeds. An error in map reading, plus flying on a further five minutes or so while checking your position, can easily take you into Controlled Airspace leading to a near miss with passenger aircraft and the likelihood of a fine of many hundreds of pounds.

In addition to the weather hazards, aero engines are notorious for carburettor icing. This can occur at any time the humidity is high and almost regardless of the outside air temperature. Carburettor icing is likely, therefore, at any time near cloud base or in rain and it is a common cause of accidents and forced landings.

If you are an experienced glider pilot starting to fly powered aircraft, never forget that the shock of impact goes up with the square of the speed. Flying into high ground or even landing on a really rough field can destroy a powered aircraft, whereas you might well survive if you are flying a microlight or a glider with its much lower landing speed and little or no risk of fire.

Altimeter settings

The normal altimeter is an aneroid barometer which is calibrated in heights instead of indicating the pressure in millibars or inches of mercury. More recently electronic altimeters have become available and these detect changes in pressure by transducers and indicate the height with a liquid crystal display reading to the nearest foot or metre. Larger powered aircraft use a Radio Altimeter for getting accurate heights above the ground and these are the only types which give accurate readings regardless of the terrain below.

Since the atmospheric pressure decreases regularly with height, pressure altimeters are a very satisfactory way of determining heights. But, of course, the ground is not level and the altimeter only gives heights relative to the level at which it has been set. If it is set at the sea level pressure it will indicate heights above sea level – providing that the pressure does not change because of the weather.

If the pressure changes during the course of a flight, maintaining a constant reading of the altimeter would result in a gradual change in height. Variations of pressure are usually fairly small during soaring weather but it is important to realise that on a long flight in a light aircraft, flying from a high pressure area into lower pressure results in worsening weather, as well as flying lower for a given indicated height. This can be remembered by the rhyme 'high to low, look below'. This is a common cause of aircraft flying into high ground in cloudy weather or at night, while according to the altimeter the pilot is maintaining a safe height.

Note that a change in pressure of 1mb is equivalent to a change in height of 30 feet. So, an error of 600 feet is possible if the pressure drops by only 20mbs. This kind of change could easily occur during a long cross-country flight in a powered aircraft flying into poor weather.

For practical purposes the effects of temperature changes cause only minor errors to the readings of an altimeter. However, serious errors occur by using the wrong subscale setting and this has been the cause of many bad-weather accidents.

In gliders, the lack of vibration often causes the altimeter to lag by several hundred feet when climbing or descending. To obtain a correct reading it may be necessary to vibrate the instrument panel by tapping it. It is NEVER safe to rely on it when landing in a field as the height of the ground will be unknown and may be very different from the point of take off.

Flight levels

To avoid possible collisions on the airways there is a common setting for all aircraft using the airways. Using this standard setting ensures that separation between opposing traffic is maintained with a safe margin. The **flight levels** used for flights along the airways are indicated heights with the altimeter setting of *1013mbs*. Of course gliders and light aircraft are not permitted to fly along the airways but it is important to be able to tell when you have climbed up close to the airway base.

The height of the base of an airway marked on the map as F.L.45 will be at 4500 feet when the altimeter is set at 1013mbs on the sub-scale. This may be a considerable difference from the height shown when the altimeter has been set at either sea level pressure (known as the QNH) or when the altimeter has been set to read zero at an airfield level (known as the QFE).

[The Q code is a shorthand devised for use with the morse code to save time repeating a lengthy explanation for common phrases. QFE and QNH are still used with speech transmissions, and there are many more abbreviations covering a vast list of common phrases, but these are seldom used by pilots.]

A simple way of remembering these two is that 'E' is for 'Easy'. QFE is the setting for an airfield which gives the altimeter a reading of zero feet on landing at that airfield. The QFE cannot be used if the airfield is at a high altitude as the normal altimeter cannot be wound down to read zero over such a large range.

'H' is for 'Harder'. QNH is the altimeter millibar setting for a location which gives the height above sea level at that location.

With this setting you must subtract the height of the ground below if you want to know your height above a particular airfield, etc. It is particularly important whenever it is necessary to know your height relative to high ground or an obstruction such as a television mast marked on your maps.

If you are flying a non-radio aircraft you can avoid calculations in the air if you make a note of the QFE and the QNH before take-off. Wind the altimeter back to read zero and note the sub-scale reading for the QFE and the QNH (by setting the height of your take off point above sea level and noting the sub-scale reading). Then at any time in the air you can wind on 1013mbs to check your height relative to the airway. A moment later you can reset the other reading to carry on with the flight. Winding back onto the QFE may eliminate the risk of miscalculating the height for a final glide back into your airfield. Nowadays most light aircraft carry radio, and airfield controllers always give the QFE with landing instructions.

Example

Here is an example of the kind of calculation that a pilot might need to do in the air if the QFE and QNH have not been noted before take off.

I am flying across the line of an airway marked F.L.45, at an indicated height of 4000 feet and with the altimeter set to the QNH of 1008mbs.

Am I in the airway or can I continue climbing higher?

The answer and calculation are on the next page, but don't look them up until you have tried to solve the problem first.

Remember, a few moments spent noting the altimeter settings before take off can save the need for such computations in the air.

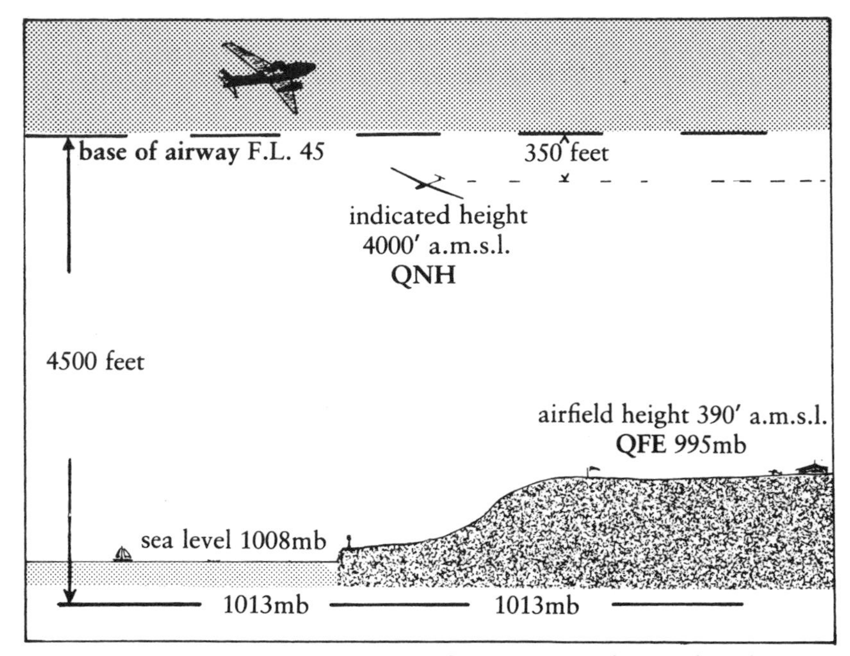

Fig 41 Altimeter settings. The pilot flying at 4000 feet on his altimeter may be very close to the airway at flight level 45.

Answer to question on page 88:

Indicated height 4000 feet with QNH 1008mbs
Airways altimeter setting 1013mbs
Difference in pressure 5mbs

5mbs at 30 feet per mb = 150 feet

Therefore, I am only 350 feet below base of the airway F.L. 45 which is at 4350 feet indicated on QNH.

If the QNH had been 998mbs, I would have been on the lower edge of the airway when the altimeter read only 4050 feet.

Note that taking off from the airfield (*see fig. 41*) with the altimeter set to QFE the base of the airway would have been at 3960 feet indicated.

(See also fig 41.)

REVISION QUESTIONS

1. What is the effect on the soaring conditions when the prevailing wind is from the sea?
2. The altimeter uses the change in pressure with height to operate the instrument. Why is the pressure less at height than at sea level?
3. Because of the rotation of the earth the air in a low pressure region moves around the centre with the winds at about 2000 feet blowing along the isobars. Which direction is the rotation in a high pressure system in the northern hemisphere?
4. What is an isobar?
5. What is Buys-Ballot's Law?
6. You are facing north with the wind blowing in your face. In which direction is the lower pressure area?
7. The wind is forecast to veer later in the day and is at present due west. Which direction will it be this evening?
8. What is the gradient wind?
9. What are Katabatic and Anabatic winds?
10. Steep wind gradients are a hazard to all landing aircraft. In what conditions is a severe gradient most likely?
11. A halo round the sun or moon is usually a sign of approaching bad weather. What kind of weather would you expect during the following 12 hours?
12. What is the warm sector?
13. What change in wind direction can be expected with the passage of a warm front?
14. What change in wind direction can be expected with the passage of a cold front?
15. What types of lift may be found in the warm sector?
16. Why should gliders be put away if an active cold front is due?
17. What often occurs on the day after the passage of a cold front?
18. What effect does the slow subsidence of air in a high pressure system have on cloud formation and soaring conditions?
19. What is likely to occur if the forecaster reports that a wave appears to be forming on the cold front?
20. What general weather is associated with a depression?
21. What general types of cloud are associated with a warm front?
22. What is an inversion?
23. Why do inversions form in a high pressure system?

24 An intensifying anticyclone results in hot sunny weather. Why is it unfavourable for soaring?
25 What is a trough of low pressure?
26 What is a ridge of high pressure?
27 What is a col?
28 A persistent high pressure system over the Continent may block the progress of Atlantic depressions. What happens to the depressions affected?
29 What effect does the freezing level have on the risk of showers developing?
30 Why is it dangerous to be caught flying in a snow shower?
31 What hazards occur in and near to showers?
32 Is water vapour visible?
33 Is water vapour lighter or heavier than dry air?
34 What is cloud?
35 What is advection fog?
36 When is radiation fog likely to form?
37 What name is given to low cloud lying on high ground?
38 What is the dew point?
39 What is an adiabatic process?
40 What is meant by the Dry Adiabatic Lapse Rate?
41 What is the DALR?
42 What extra heat is released when cloud is formed?
43 What is the Saturated Adiabatic Lapse Rate?
44 What is the composition of the anvil of a cumulonimbus cloud?
45 VOLMET gives the temperature 19°C and the dew point 11°C. At approximately what height would you expect the base of cumulous clouds to be?
46 What is a supercooled water droplet?
47 At what height is stratus cloud usually formed?
48 What is a lenticular cloud?
49 What happens to the base of the cumulus clouds during a summer morning?
50 At 6 p.m. the VOLMET gives a report for Gatwick as 'temperature 15°C, dew point 13°C. What is likely to happen to the weather?
51 What will be the effect of the sea breeze near a coastline if the prevailing wind is onshore?
52 How can you recognise a sea breeze front?

53 What happens to the thermal activity on the sea side of a sea breeze front?
54 What is orographic cloud?
55 At what time of year is a sea breeze front likely to penetrate inland more than 20 miles?
56 Name three good sources for thermal formation.
57 When is the risk of over-development of the cumulus clouds increased?
58 What conditions of wind and stability of the air are most conducive to the formation of lee waves?
59 Describe rotor cloud?
60 Where is the best lift in relation to a lenticular cloud?
61 Is it possible to soar above a lenticular cloud?
62 What usually happens if you keep circling in wave lift?
63 Where is the turbulence in the rotor likely to be least?
64 What is the best action to take if you fly into an area of strong sinking air in the lee of a range of hills?
65 What happens if you fly behind the hilltop when ridge soaring?
66 What conditions result in greatly improved hill lift?
67 What is a QFE?
68 What is a QNH?
69 What is the altimeter setting used for flight levels in an airway?
70 Flying into bad weather, what error could you expect on the altimeter?
71 The QNH is 1009mbs and the airfield is 330 feet above sea level; what should the QFE read if the altimeter is accurate?

Answers

1 The stable sea air penetrates inland and spoils the thermal activity, often after a promising start to the day.
2 There is less weight of air above than at sea level.
3 Clockwise.
4 A line joining places with equal sea level equivalent pressures.
5 The Law states that with your back to the wind, the low pressure is always to your left (in the northern hemisphere).
6 To the east or to the right of you.
7 Northwest.
8 The wind velocity at 2000 feet (above the friction level).
9 A Katabatic wind is the result of the air on hills or mountain tops cooling down, becoming more dense and flowing down

the slopes into the valley. An Anabatic wind is a shallow upward flow of air close to the hillsides caused by them warming up more quickly than the valley.

10 Strong winds and rough surfaces or obstructions upwind of the landing area.

11 A gradual thickening of the upper cloud and lowering of the cloud base, a strengthening and backing of the wind and then rain (i.e. it is a sign of the approach of a warm front).

12 The area between a warm and cold front in a depression.

13 Wind veering, usually from south to southwest or west.

14 Wind veering, usually from the west or southwest to northwest.

15 Possible for lee waves but poor for thermals.

16 An active cold front may involve thunderstorms, hail or very heavy rain and squally winds with severe gusts. Tyres, etc, on the wingtip may be ineffective because of the sudden large changes in wind direction on some fronts.

17 It is a super soaring day.

18 The subsidence tends to limit the cloud development and prevent over-development and cycling. A persistent high pressure system will result in deteriorating thermalling conditions because of the inversions formed.

19 There will be bad weather with strong to gale winds.

20 Strong winds, cloudy conditions and precipitation.

21 Mainly layer clouds such as nimbostratus and stratus.

22 A situation where the temperature increases with height.

23 The subsidence of the air causes an increase in temperature due to compression at medium heights but not at low levels. In addition, the clear skies at night result in rapid cooling of the ground and the air immediately above it. The temperature therefore increases with height in the lower layers.

24 The air is too stable and the inversions inhibit the thermal activity.

25 A rather elongated area of low pressure without any distinctive fronts but usually bringing rain.

26 A narrow wedge shaped area of high pressure between two areas of low pressure.

27 An area of slack pressure between two depressions and two areas of higher pressure.

28 They usually move north affecting Scotland and the Scandinavian countries.

29 The lower the freezing level, the more likely for showers to develop.

30 The visibility can be reduced to almost zero in a few minutes.

31 Sudden changes in wind direction and strength, severe icing, hail, lightning and very strong up and downdrafts.

32 No.

33 Lighter.

34 Visible water droplets.

35 Fog formed when moist air is cooled by flowing over a cool surface such as the sea.

36 In a high pressure system when there are clear skies and light winds at night.

37 Hill fog.

38 The temperature to which the air must be cooled down to become saturated.

39 A process where no heat enters or leaves the system and where the temperature change is solely due to changes in pressure.

40 The regular decrease in temperature with height of dry air, caused by the expansion of the air as pressure is reduced, assuming no other cause for a loss of heat.

41 1°C per 100 metres.

42 The latent heat used to evaporate the moisture on the ground and turn it into water vapour.

43 About 0.5 degrees per 100 metres.

44 Ice crystals.

45 About 3200 feet (400 times the difference between the temperature and dew point in degrees C).

46 An unfrozen droplet of water with a temperature well below the normal freezing point of water.

47 Below a thousand feet, often covering the hills.

48 A bar of cloud, sometimes lens shaped, formed by wave activity and lying parallel to the hills producing the wave.

49 It rises rapidly.

50 Risk of fog forming as the temperature is almost certain to drop more than 2 degrees during the night.

51 There will be little or no thermal activity.

52 By the double cloud base and the line of curtain cloud or, if no cloud, by the sudden drop in visibility and change in the surface wind.

53 Thermal activity dies out and then later becomes weak with turbulent, broken thermals.

54 Cloud formed when the air is lifted by flowing over high ground.
55 Summer.
56 High ground and particularly sun-facing slopes. The upwind edges of lakes. Towns. Airfields. Stubble fires.
57 When there is an inversion or stable layer limiting the height of the cloud tops and the air at that level is moist. With the approach of a depression.
58 Surface wind at hill top level of at least 20 knots and at right angles to the ridge of hills. An increase in wind speed with height but with a constant direction. An unstable layer up to about the height of the hills with a stable layer above it.
59 A stationary line of broken cumulus clouds with a torn appearance and obvious swirling motion lying just downwind of a ridge of hills or mountains and below the lowest lenticular cloud.
60 Just upwind of the edge of the cloud.
61 Yes, the cloud forms at the level where the air becomes saturated by being lifted and cooled and does not indicate the top of the wave system.
62 You drift back into the strong sinking air within a very short time.
63 At the ends of the wave systems and wherever the wave lift is weakest.
64 Turn downwind. Do not fly parallel with the ridge while in sinking air.
65 You fly into the 'curl over' and lose height very rapidly onto the hilltop.
66 Thermal activity and the passage of a cumulus cloud over the ridge, and wave lift if the hill becomes in phase with a wave system from another hill up wind.
67 QFE is the altimeter millibar scale setting for an airfield which gives a reading on the altimeter of zero on the ground at that airfield.
68 QNH is the altimeter millibar scale setting for an airfield which gives the height above sea level, of that airfield.
69 1013mbs.
70 Over-read. Remember 'high to low, look below'.
71 998mbs. 330 feet at 30 feet per millibar reduction in pressure is 11mbs. The airfield height of 330 feet is indicated with the subscale at 1009mbs, so zero feet will be indicated with it at 998mbs.

INDEX